BUILDING EGYPTIAN PYRAMIDS
Achieving the Impossible

Szeamus Chapman

Revised 2016 for Kindle Textbook format to include additional text, diagrams, colour photographs and video

© 1998-2016 Seamus Chapman
All Rights Reserved.

1 Royal Cubit = 53.4 centimetres = 20.6 inches

CONTENTS

PART I

Fundamentals

Introduction

Background to this enigmatic puzzle. A perspective of the volumes and type of materials and timetables. Comparison with modern production rates both mechanised and manual.

Pyramid Construction Fundamentals

Identifying by examples the fundamental processes necessary for making any pyramid accurately. These fundamentals when applied to a massive pyramid.

Controlling the shape of any Pyramid

Delivering Materials

Alternative Proposals

A summary of how others have suggested large pyramids might have been made. Do they identify and resolve, or ignore the fundamentals identified?

Controlling the Shape

Delivering Materials

Pyramid Dimensions and Shapes

The accuracy of pyramid dimensions.

Alignment of Pyramids

Why and how pyramids were aligned to the cardinal points.

Preliminary Conclusions

Pyramid Corner Edges and the Virtual Apex

The geometry of symmetrical pyramids. How this can be applied to provide the corner edges of a pyramid without the apex for reference.

Reference Squares and the Virtual Centrepoint.

How the side or diagonal dimension of certain squares provides a simple method for calculating their centrepoint-corner dimension.

Forming any square with a known dimension to its virtual centrepoint, even when this is covered.

An alternative method of forming a right angle and partial diagonal in any square.

Discovery in Ancient Egypt

Background to this discovery.

Evidence in Ancient Egyptian measures.

Access Ramps and Platforms for the Delivery and Fitting of Materials

Description of a ramp/platform system, which provides external access to a pyramid at each course level, and maintains the same arrangement to the apex. How its cross section geometry exploits the shape of the pyramid

Examples of Major Egyptian Pyramids

Their diagonal cross-section dimensions

Their height : centrepoint – corner ratios

How they were determined and lead to the profiles found.

The Southern (Bent) Pyramid

The Northern (Red) Pyramid

The Meidum Pyramid

The Great Pyramid of Khufu

Khafre's Pyramid

Menkaure's Pyramid

Height : centrepoint - corner ratios

Geometry showing corner edge proportions leading to profile angles of Egyptian pyramids

The layout of the Giza Plateau

Conclusions

PART II
Building the Great Pyramid of Khufu

Details of all aspects of the construction of this pyramid, including the preparation of typical materials, how the virtual apex, virtual centrepoint and ramp/platform system, provide the vital elements in forming an accurate pyramid shape in solid stone and within a given time frame.

Preparation of the Site and Materials

Quarrying stone

Stone for the Core. Backing Blocks and Internal Features.

Stone for the Casing

Cutting Stone

Limestone

Granite

Prefabrication of Blocks for the Facade

Levelling and Marking the Base

The Pavement Layer

Delivering and Fitting the Base Course

The Reference Core

Choosing the first reference height

Placing the centrepoint and pyramid base corners

Fitting the base course

The perpendicular ramp

The Façade to the first Reference Height

Levelling the top of the Base Course

Marking the Corner Edge and the Pyramid Faces

Courses above the Base Course

The external platform/ramp system

Completing the pyramid to the Apex

Alternating Stages

Cutting the Pyramid faces

Internal Features

The Descending Passage and Subterranean Chamber

The Ascending Passage

The Queens Chamber and Grand Gallery

The King's Chamber

The Relieving Ceilings

The Air Shafts

A Change in Plan?

Installation of Passages and Chambers

Timetables and Workforce

The Core to the first Reference Height

The Façade to the first Reference Height

Workers, Effort and Time

Completing the Pyramid

Quarrying and Preparation of Material

Conclusions

Comments, which support the skill and effort of the Ancient Egyptians in creating markers of their technology, repeated throughout history, but in different ways.

Construction techniques and processes which continue to be used today; some re-discovered.

A summary of the versatility of the virtual apex and virtual centrepoint methods and how they could be applied to build large pyramids of any shape, accurately.

How the Ancient Egyptians were forced by their limited mathematics to exploit these methods in their simplest form.

Supplements

The coincidental appearance of Pi in the dimensions of the Great Pyramid

Building a pyramid using geometry only.

The advantages and evidence of two-stage construction

Course Height Patterns in the Great Pyramid

Adjustments to the Eiffel Tower

References, Sources and Acknowledgements

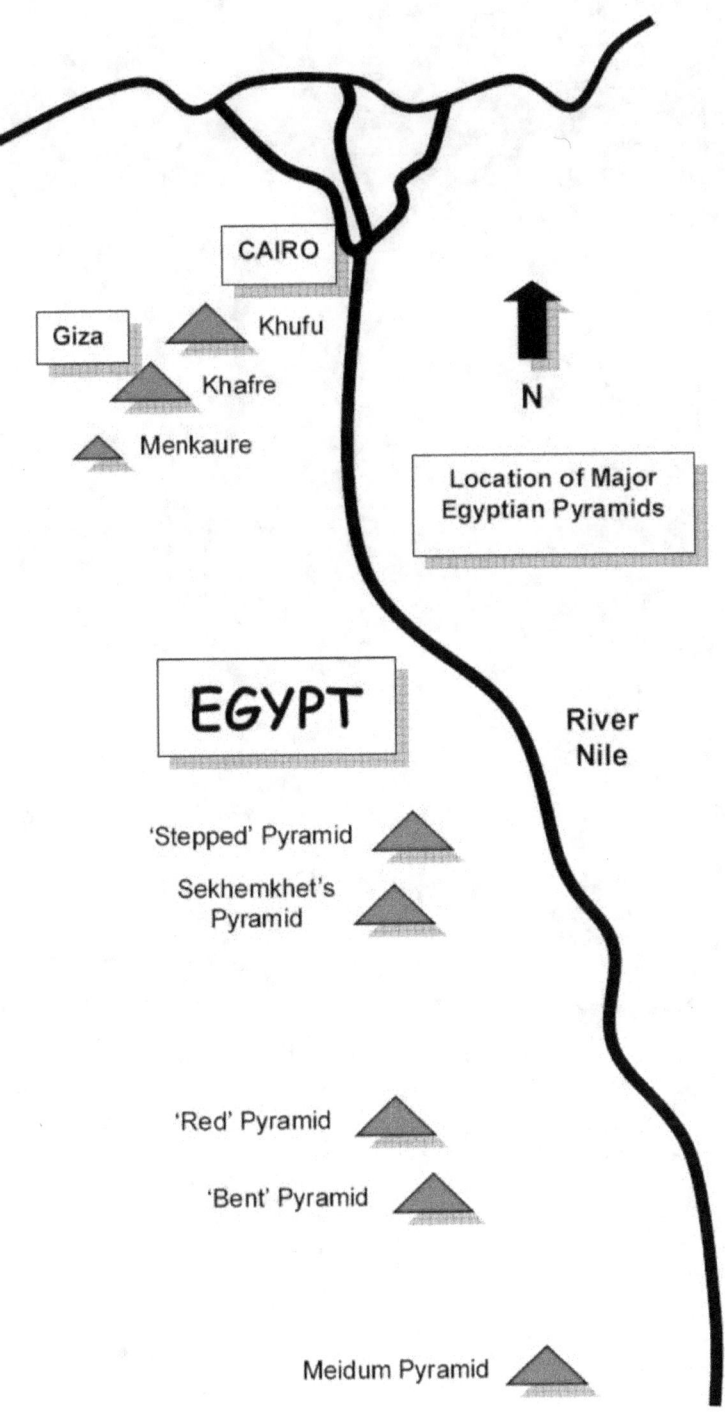

Fundamentals

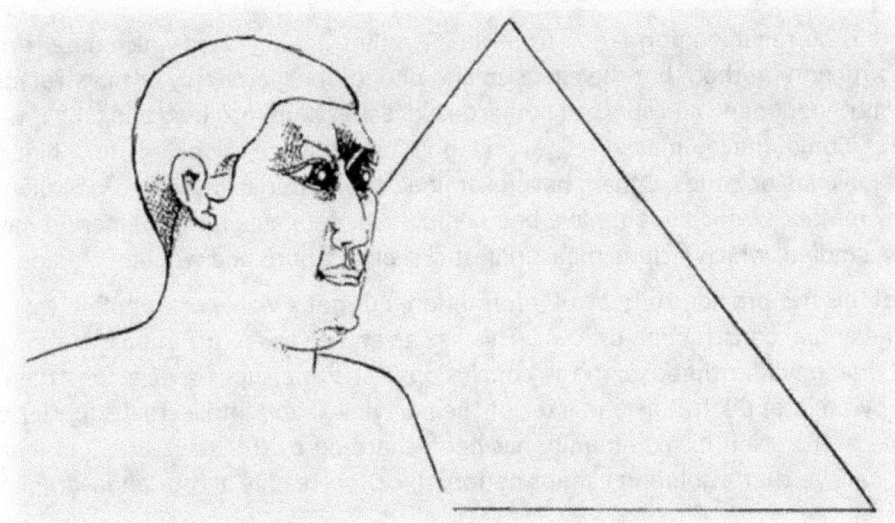

Hemon, Vizier to Khufu – Builder of the Great Pyramid

Introduction

The Great Pyramid at Giza is the only remaining Wonder of the Ancient World. It is the most distinctive and recognizable symbol of a culture which flourished over forty centuries ago.

Many man-made structures provoke awe and admiration; the Parthenon in Athens, the Coliseum in Rome, St. Paul's Cathedral in London, the Blue Mosque in Istanbul and the Great Wall of China. All of these are enormously impressive examples of human achievement.

The Great Pyramid of Khufu at Giza is unique in that over forty centuries after its construction, Scientists, Egyptologists, Engineers, Mathematicians and Architects cannot agree about how exactly it was made. Like many other aspects of Ancient Egypt, it is not readily yielding of its mysteries. If it did not exist for all to observe, measure, photograph and analyse, there would almost certainly be a body of opinion to maintain that this great wonder could not have been constructed during that period of history, or that its dimensions had been exaggerated. But the Great Pyramid at Giza is resplendently real, just a bus ride from Cairo centre, and invites us to theorise.

The statistics of its construction are impressive. It is generally agreed that it took around 20 years to build. In this period, a solid stone structure of 2,600,000 cubic metres (over 9 million cu. ft.) was created with exterior stonework of incredible accuracy. The pyramid base covers over 50000 sq.m (over 12 acres) and it has a height of almost 150 metres (over 480 ft.), with all of this achieved without the use of anything which we would recognise as a machine.

There is no reliable information from the Egyptian Old Kingdom which details the construction method, but the age, size and phenomenal accuracy of the structure has intrigued many and sparked numerous investigations into how and why it was built. Some have measured every possible feature, looking for hidden relationships or codes. Others have examined the materials used and postulated likely methods which might have been employed. Artefacts from the period have been studied, which include measuring sticks, plumb bobs and wooden cradles.

All of this has provided a body of information, but not a viable or complete theory of the actual construction process. This has given rise to an 'it's impossible' myth and driven wilder theorists to the conclusion that some super-specie constructed the pyramid and left it as a marker of their technical and intellectual superiority. Some of the scientific community has been charmed by this structure, and many half believe that a solution cannot be found, or too readily accept an incomplete answer.

Latterly this topic has attracted all kinds of theories, which exploit this lack of a complete and practical solution, and allowed these authors to introduce ideas, which have no basis in either fact or logic and at the very least add confusion to the topic.

Some proposals for example, have suggested that pyramids incorporate star alignments, which represent key elements of an Ancient Egyptian religion, but without explaining how these supposed features were actually formed within the pyramid. Others have begun with an assumption that all blocks are of equal size, and simply investigated the ways that these 'average' blocks might have been moved. In fact the visible blocks on the exterior of the Great Pyramid for example, vary in weight from 20 tonnes at the base to less than one tonne in the majority of courses and even these are not representative of the whole structure.

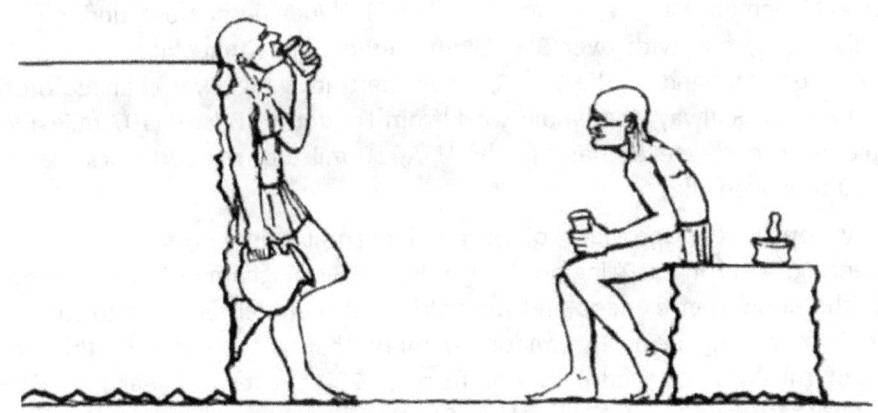

**Great Pyramid Casing Blocks
The tallest and the average**

The Great Pyramid today has had its smooth white limestone casing removed, leaving stones which were behind the casing, exposed as steps at each course level. The unromantic truth about this is that the citizens of early Cairo simply took the white casing stone for other building projects. All pyramids became a ready-made quarry, and the stone removed, was used to build the Mosques, bridges and walls of Cairo.

The statistics relating to the movement of material should provide a perspective for understanding what might have taken place. To complete the Great Pyramid in 4000 days for example, would have meant delivering and placing each day 650 cubic metres of stone weighing 1700 tonnes. Simultaneously 25 metres of casing would have to be laid so accurately that all would eventually meet at a single point at the apex, almost 150 metres above the base.

However, some modern examples of average work rates are also impressive.

During World War II, Liberty Ships weighing thousands of tonnes were completed at the rate of one ship every three days. A Toyota factory in Japan produces 1600 cars every day or 1 car per minute, even though they have outdoor storage for only 200 cars. Their total domestic vehicle production passed 100 million in 1999. Assembling the Empire State Building from steel sections prefabricated 300 miles away, was at the rate of one storey per day and took less time to complete than the predicted 18 months. It is reported that some of the steelwork was still warm when it arrived on site.

It is obvious that to produce the results in each of these examples, an organised and appropriate production system had been prepared and was being applied.

Other achievements and without machinery, include Vietnamese underground guerrilla bases; one with over 350 kilometres of interconnecting tunnels (220 miles), all dug by hand. In the 19th C, when the track gauge was changed on the Great Western Railway, the whole work from London to Bristol (107 miles) was completed in one weekend and in the USA, 10 miles of railroad track was laid from scratch in one day.

Railway construction provides other good examples in non-mechanised civil engineering. A cutting at Tring on the London to Birmingham line is 4 kilometres in length and 20 metres deep and the volume of material removed to form the Chalk Farm cutting in north London, is more than the volume of the Great Pyramid. This was achieved by a workforce of 20000 over a 5 year period and their work rate tallies with that of a 'good navvy' from the period who would be expected to move up to 20 tonnes of 'muck' up and into a wagon each day.

The Great Pyramid is the largest and most well-known of all the pyramids, but others constructed during the same period are also large and accurate, even though each has a slightly different shape. It follows that any solution to the mystery of the construction processes employed in the Great Pyramid must also apply to all others and clearly demonstrate how a work rate can be achieved, with the materials and tools known, to create the structures with the shapes found. To focus only on the moving of large blocks for example is not enough as this aspect is only one of the many inter-related processes in the completion of the design.

It is necessary to show, not only how the Ancient Egyptians might have moved typical blocks of stone, but what was done with them when they arrived at a pyramid and why.

Pyramid Construction Fundamentals

The starting point in any design process is to identify the fundamentals of the design brief. In the case of a pyramid, an appropriate starting point is to consider how any solid pyramid might be made.

For example, to make a small pyramid of say 1 metre high, with sides of length 110cms, it is relatively simple. Mark a square base of 110cms. and from the corners, using the diagonals, find the centrepoint. A vertical pole 1 metre high placed on the centrepoint would determine the apex to which the corner edges could be aligned using a straightedge. From these accurately positioned edges, the face slopes would be automatically formed, by taking straightedges from corner to corner at various heights.

Another example is to consider how the capstone of a pyramid might be made to the same dimensions. In this case a mason would begin by carving a flat base and marking the corners of a square with sides of 110 cms. After inverting the block onto its level base, he would transfer the base corner positions to the top, using vertical channels running up from each corner to a height of 1m. On the levelled top the centre-point can be found from diagonals connecting the marked corners. Four straight channels would then be carved, which connect this centre-point to the base corners and define the corner edges and apex. Surplus stone is then removed from each face until a straightedge confirms the flatness of the faces between the corner edges. Each of these examples is simple, accurate, would be used today and with no credible alternative.

In both examples the apex was used to find the pyramid corner edges and from these the pyramid sides. In each case straightedges connected these primary reference points.

It is at this point that any alternative methods of making even a small pyramid accurately should be tested, with particular attention paid to any elements requiring trial and error.

It can be seen from these examples that Egyptian pyramids present particular problems, because their great height meant the apex was not in place until the final stages of construction and therefore not available as a reference point.

Also, in the case of some, possibly all Egyptian pyramids, the centre of their bases was left un-cleared, thus preventing the use of diagonals to either confirm the accuracy of the square or to position its centrepoint.

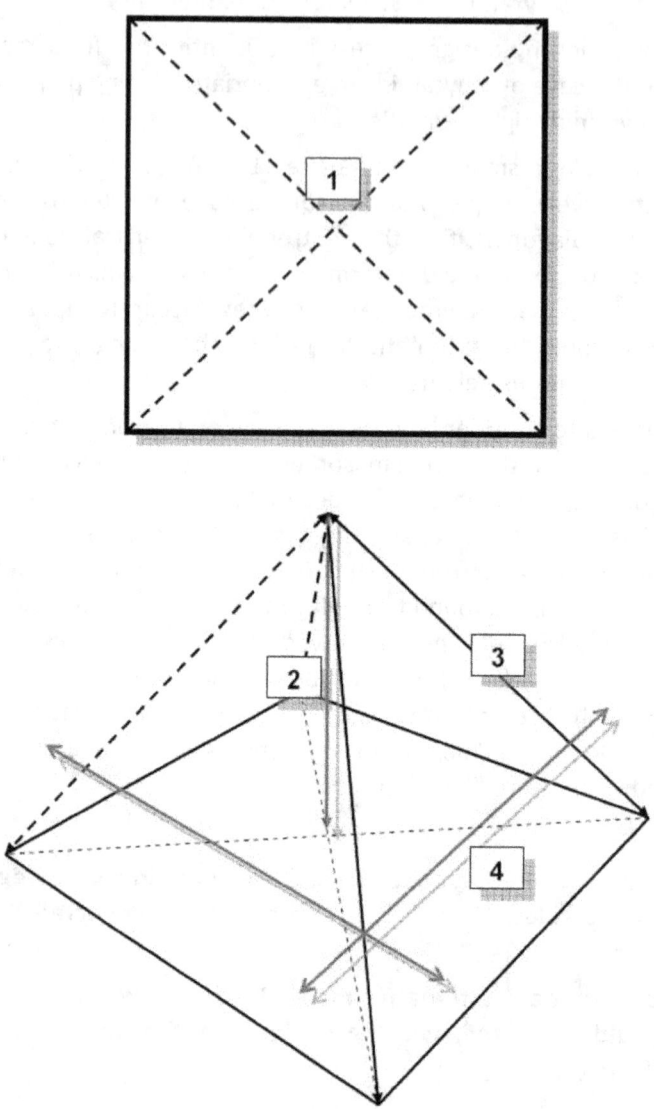

Building a small solid Pyramid

1. Form a square and find the centrepoint using diagonals.
2. Fix a vertical marker at the centrepoint of length equal to the chosen height.
3. Form the corner edges using a straightedge to the apex.
4. Form the faces using a straightedge between adjacent corners.

What alternative methods might the Ancient Egyptians have employed, which must provide each of the fundamental alignment features identified and necessary to make their pyramids accurately?

It is clear from both the above and from normal building practice that the placing of corner positions at various heights would be vital for providing a straight line to accurately position the blocks which form the pyramid sides and ultimately define the face slope. In this case there is a need to identify how this primary reference can be provided to meet this fundamental aim.

How did the Ancient Egyptians know where the pyramid corners should be at every course height, before the apex was in place?

Another fundamental which must be resolved, is how were the materials delivered to structures of such great height?

Egyptian pyramids in some cases are built from millions of tonnes of stone and are over 100m tall. This problem can be summarised in the following way:

How did the Ancient Egyptians deliver and fit the capstone on a pyramid?

Related questions must also be answered.

How did the Ancient Egyptians align their pyramids to the cardinal points and how did they make large squares accurately, without using diagonals yet with a known dimension from the centrepoint?

Alternative Construction Methods for Controlling the Pyramid Shape

Other pyramid construction methods have been presented which attempt to explain how some of these problems might have been resolved.

One of these methods is based on building a pyramid to a profile which has a specific height : half base ratio.

For example the Great Pyramid which has a height of 280 Royal Cubits and a base side length of 440 Royal Cubits, is described as having a height : half base ratio of 14:11. Effort has therefore been focussed on suggesting ways in which this ratio – or the Ancient Egyptian equivalent - could have been used directly as the primary reference in forming the pyramid shape.

'Rise and Run' is a method, which proposes that each exterior casing block is marked with a face slope, based on its chosen profile. When these blocks are carefully placed on a level and straight course of blocks below, this would allow the faces of the pyramid to rise to the chosen height, with all four sides and corners eventually meeting at a single point at the top. However, blocks from pyramids which have had their slope angle measured on a level surface have been found to differ from their pyramid slope to such an extent that some would also fit on other pyramids with a different slope.

It is suggested that these deviations in slope angles, plus or minus, would be automatically 'self-correcting overall' during construction. As some pyramids also have casing stones laid at an inclined angle, the face slope of each of these blocks would therefore have to have been the chosen pyramid slope angle minus the incline angle.

Tilted casing courses on the 'Bent' Pyramid

The pyramid below has the blocks forming the sides of the façade laid on an incline, while the corner blocks are laid on a level base.

Djedefre's Pyramid at Abu Rawash

An alternative method of applying 'Rise and Run' is for the builders to check the pyramid face slope during construction by measuring inwards at specific heights, an amount which corresponds to the height : half base ratio.

Backsighting is proposed as an additional aid which suggests that the pyramid shape is controlled by visually comparing blocks as they are laid with those already in place, with an aim of ensuring that the pyramid faces and corner edges follow straight lines. A variation of this method which attempts to overcome obvious weaknesses, proposes an internal core structure is constructed first with a pole eventually planted on top at the apex position, thus providing a single point to which the corner edges and sides of the casing can also be visually aligned. For the Great Pyramid it would be like sighting to the top of a 40 storey building, even assuming the pole is in the correct position.

However visual sighting does not provide any help in constructing accurately and at best would only provide a means for identifying errors after the events, which are then impossible to correct.

None of these methods explain how the straightness of a course of blocks is maintained when there is no physical guide. Each wrongly assumes that a pyramid shape can be accurately formed by carefully placing marked blocks, which might already contain errors and without an opportunity for making adjustments until later. They would have to rely entirely on the straightness of an earlier course of blocks as the only guide for placing the blocks of a subsequent course. In this case the straightness of hundreds of courses of casing blocks could only have been derived from the base course, which was uniquely laid to a stringline. If the base square were accurately formed in this way, why would the Ancient Egyptians abandon this reliable and universal tool?

A skilled bricklayer for example, would find it impossible to build a wall of any kind using visual alignment only to keep it straight and vertical.

Instead he would first provide accurate corners before connecting them with a stringline which when stretched would define both the straightness and the perpendicular of the section between.

Even in the case of a building requiring a rounded corner, conventional reference corners are built first in order to provide a means to accurately align the straight sections of the adjacent walls up to the point at which the curved section begins. The two ends are then connected by the chosen radius with the reference corners demolished leaving no trace of their presence.

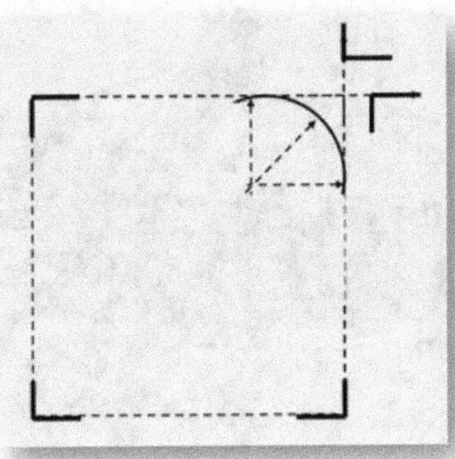

If it were in fact possible to build a large solid pyramid using any of these 'rise and run' techniques then it would have been more reasonable for the Ancient Egyptians to have simply cut or marked all their casing blocks with vertical faces and then tilted the course so that the face matched the desired 'rise and run' angle. However in all the proposed applications of 'rise and run' methods including this one, there is an issue of when the casing blocks are placed.

If the casing is placed last then the exact depth to its backing block must also be accurately prepared to receive it if the shape is to be maintained. This is despite each casing block being of a random depth. If the backing blocks are placed after the casing is installed, then how are they taken behind them and brought routinely into position?

Building other structures of great height with a sloping profile

Tall chimney construction in the 18th and 19th centuries provides a consistent example of how to form a vertically accurate structure when a 'batter' is included. The alignment references used throughout construction were taken from the centre of the shaft; its vertical position routinely checked using a plumb line. The 'batter' was formed by steadily reducing the outside diameter according to a simple formula based on brick dimensions and the radius.

Assembling the Eiffel Tower from pre-fabricated components

Leg at base - 1 of 4

Construction of the Eiffel Tower clearly illustrates the difficulties of inclining corners accurately when there is no physical reference. The designers used the tower's elevations as the basis for the design and referred to their scale drawings to prefabricate all the parts to an accuracy of 1 millimetre. As each of the 4 legs were assembled, they had to be inclined to a 2:1 ratio in two directions simultaneously – a compound ratio of 2:√2 – and to a height of over 50 metres, where the first platform would be placed. In order to make the connection, they used pre- installed hydraulic jacks in each leg and sand boxes on the platform, to make the necessary adjustments in three dimensions when bringing the components together.

Had this been a solid stone pyramid to the same design it would have been impossible to control the shape using a trial and error method such as this.

Pyramid Corner Edges

Significantly, not one of the methods proposed for controlling the shape of a large pyramid is able to describe how the corner edges are formed where the pyramid faces meet. Some pyramids with their casing sufficiently intact, show face slope angles which vary on each side and some have faces which are not completely flat, because none have base sides of equal length or whole Royal Cubits. However all have corner edges running straight to their apex, where an error of only one degree at the base of the tallest, would have caused them to miss by over five metres at the top.

In pyramid building the centrepoint-corner dimension for example, is an essential and vital element in controlling their shapes, but has been ignored in other construction methods, due only to an erroneous assumption that its geometry would have been too complex to apply in the Pyramid Age. As a consequence of discarding this practical alternative, attention has focussed instead on applying the height:base ratio only, causing each of the proposed methods to demand increasingly impractical and ultimately exotic ways of implementing them.

Note that: Using the base length and height proportions as an alignment method would force casing block placement and ultimately the pyramid's shape to rely primarily on visual sighting – which would require all lower sections of the structure to remain visible during construction – thus requiring complex ramps of huge volume, which can only be attached at one point only – thereby restricting the movement of materials and eventually running out of room – which must then be supplemented by deploying exotic machines, pulleys, levers, runways, steps and even tunnels - each accumulating unresolved issues of materials volume, delivery, accuracy, access, double handling, timetables and effort.

Much of the confusion over methods of solid pyramid construction has arisen because it has been assumed the problems of alignment only apply to large pyramids. This is not the case. The fundamentals of accurately forming a solid pyramid without the apex as a reference, apply to all pyramids.

Any construction method can therefore be tested on a small pyramid as the same alignment problems will be presented once the base square has been marked out and material is added in an attempt to form the shape.

A true design process when presented with such fundamental demands will continue to explore every possible alternative means of effectively achieving the given objective, before any attempt is made to put them into practice.

A process that applied in Ancient Egypt as it does today.

Delivering Materials

How millions of tonnes of stone were systematically delivered to a pyramid is one of the most intriguing aspects of their construction. Evidence of ramps is found at the base of some pyramids, but there is nothing which describes how material was taken to the highest levels. Many ramp designs have since been conceived, but as a consequence of a flawed pyramid construction process, only considered how a ramp of reasonable slope and volume might be provided. They fall into various types as each attempts to overcome the deficiencies in another, but each has ignored the need for an external platform at each course level of the pyramid.

A single perpendicular ramp can provide a reasonable slope, but if so, fails by being impossibly long and of huge volume.

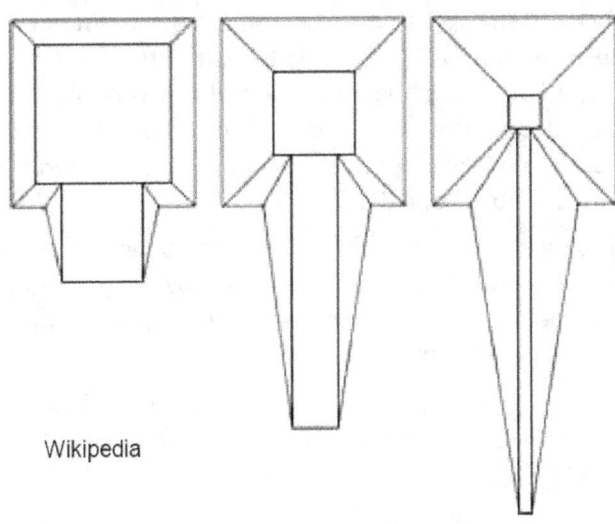

Wikipedia

External spiral ramps of various designs require less material, but each of those described runs out of room long before they reach the apex.

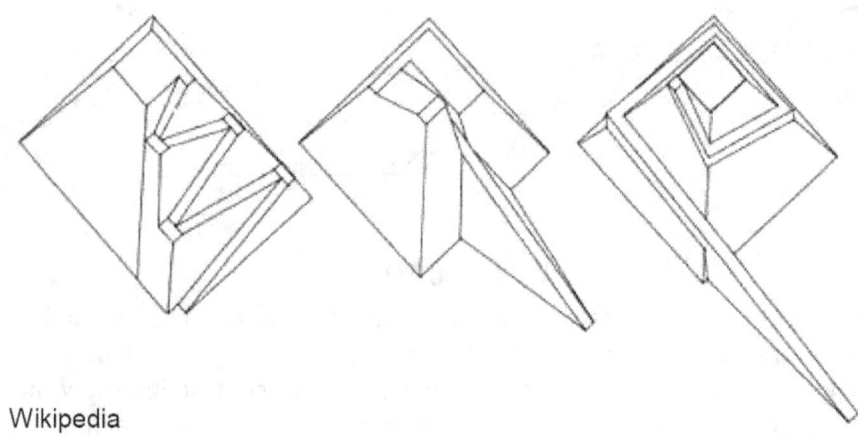

Wikipedia

A common factor is each of these ramps only arrives at a single position on one side of the pyramid at each height.

Supplementary designs include multiple perpendicular ramps, or ramps running onto and through the core. Some of these ramps connect to steps built on the core, to provide a means for the delivery and fitting of blocks at the highest levels. These also require blocks to be levered up the steps, or employ a crane of some kind. Machines have been suggested which either move or lift blocks of stone into place. They are usually specifically designed for either the most massive blocks or a so-called 'average block'. Each ignores the full range of sizes and number of blocks actually used and the pace required for their delivery. One method attempts to sidestep the delivery problem by proposing that pyramid blocks were cast in-situ, using a primitive concrete.

In a block moving race to move 1 block, 1 metre above the ground a machine might appear to be useful, but when asked to move 10,000 randomly shaped blocks to the same height, constructing a reusable ramp and dragging the blocks up it, would be far more efficient.

There is no evidence of any machine being used in Ancient Egypt and in massive solid stone pyramids they have no place. Machines, like animals also, would have simply got in the way.

Access to a pyramid can be compared with spectators entering and leaving a sports stadium. Here, 50,000 can enter or leave in less than I hour, because there are multiple points of access and egress, known to each individual. Imagine a situation if these conditions were absent, with each spectator carrying a heavy suitcase and accompanied by a dog.

The human labourer is effective in the confined workplace of a pyramid, because what he might lack in physical strength is more than compensated for in skills of communication, initiative and teamwork.

For a materials delivery system to be effective in pyramid construction, it must provide external access to the whole of each face at each new height, be of low gradient and volume and maintain these features to the apex.

Egyptian Pyramid Dimensions and Shapes

Egyptian pyramids have been surveyed and measured by many over a period of more than 100 years. The Great Pyramid is said to be the most carefully measured building on earth. The accuracy of the dimensions found and the way in which they were determined has depended on the condition of the structures themselves, but great care has been taken in each case. It can be noted however that every survey focussed only on their profile dimensions

Surveys which have been repeated independently, have confirmed earlier results and today there is general agreement about the dimensions and shape of the major structures. Many of the earliest surveys were carried out to determine if a pattern could be found in the dimensions or shapes of pyramidsThis data has been employed since in an attempt to identify the reasons why the Ancient Egyptians chose particular shapes for their pyramids.

In addition to their base and height dimensions and the angle at which their sides slope, Egyptian pyramids are also described as having sekeds.

This Egyptian term refers to their height-centre-side right triangle profiles, as a height : base ratio. The seked is based on the Royal Cubit and its 28 fractions of 7 palms of 4 fingers. The Royal Cubit is accepted as a standard measure in Ancient Egypt and examples of contemporary Cubit measuring rods have been found. For example Khufu's pyramid is said to have a seked of 1 Royal Cubit to 5 palms + 2 fingers (28:22) and Khafre's pyramid 1 RC to 5 palms + 1 finger (28:21).

Although there are no contemporary references to the seked, much later generations did employ it to describe the shape of Old Kingdom pyramids and even included it in some mathematical tests. This is one of the reasons the 'rise and run' method of forming pyramid shapes has been so readily advocated and accepted, even though in reality it does not provide an adequate means for accurately controlling the shape of a large pyramid.

It might be the case that the seked was itself devised by these later generations of Egyptians as a way of making sense of the dimensions of completed pyramids. It provided them with a method for approximating their heights when only the base dimensions and face slope were available for measurement.

If it were true that the choice of shapes for Egyptian pyramids was based only on a seked, it is surprising that the dimensions of some do not always form a clear seked ratio, despite the fineness of the scale used and the accuracy of the structures.

It can also be noted that no large pyramid has sides of equal length or a whole number of Royal Cubits and it follows that each side must have a different slope.

Preliminary Conclusions

The flaws in the content, logic and application of the methods proposed for controlling the shape of a pyramid, or delivering materials to it, demonstrate that none of them could have been successfully applied to build an Egyptian pyramid. It is quite clear that trial and error can play no part in forming the shape of a large pyramid. Eiffel's inclusion of adjustment mechanisms in the legs of his tower confirms that he understood the problem, but not how to resolve it in a simple way.

It is fundamental an accurate physical reference is available throughout construction for ensuring that the corner edges run straight, the four faces are flat and each will meet at the chosen height.

Any method describing the construction of Egyptian pyramids must begin with at least a means for accurately defining their corner edges, without the apex for reference, which when extended will meet at a single point at the chosen height – **the virtual apex.** The provision of this guide can only be derived from the pyramid corners at different heights, which must be known before the blocks of stone are placed.

It must also provide a method of accurately forming large squares without measuring across their base, but with a known centrepoint-corner dimension identifying **- the virtual centrepoint.**

A delivery system must be available which allows the transportation of blocks by human effort alone and to all parts of the pyramid, including the exterior and at every height, including the apex.

The methods must be consistent and account for the different shapes of Egyptian pyramids and all the physical evidence.

If these conditions are met, then it is more likely that these methods were used by the Ancient Egyptians, than any other.

Before embarking on building their structures, the Ancient Egyptians must have recognised and resolved these fundamental construction processes, necessary to build a large stone pyramid accurately and with the manpower and technology available to them.

The purpose of this work is to identify and resolve the same fundamentals and demonstrate that when applied, the Great Pyramid and all others could have been constructed using cleverly applied geometry, primitive technology and large amounts of effort and ingenuity.

Alignment of Egyptian Pyramids

The Merket and Bay

It is reported that all Egyptian pyramids have one side running northwards and as they are square structures their other sides run to the other three cardinal points. It is certainly helpful in forming large squares without using diagonals, to ensure that at least two sides are parallel and aligning to a distant stellar reference might have been an attempt to achieve this.

However those descriptions of methods which might have been used are exclusively based on sighting to a northerly star or groups of stars. In each case the procedures described are both clumsy and unreliable, as they involve the use of artificial horizons or plumb lines, with the northerly alignment based on bisecting the change in position of a star or stars.

All of this activity taking place at night by surveyors working hundreds of metres apart.

Haarck and Thurston both suggest that Egyptian pyramids might have been aligned eastwards to the rising position of the binary star beta Scorpii. They compared the easterly axis of the major pyramids and found that with the exception of Khafre's pyramid, each was aligned exactly to the rising position of this star according to their accepted order and dates of construction. The differences from due east found in individual pyramids followed the effects of precession, with beta Scorpii coincidentally rising almost exactly due east, when the Great Pyramid was constructed.

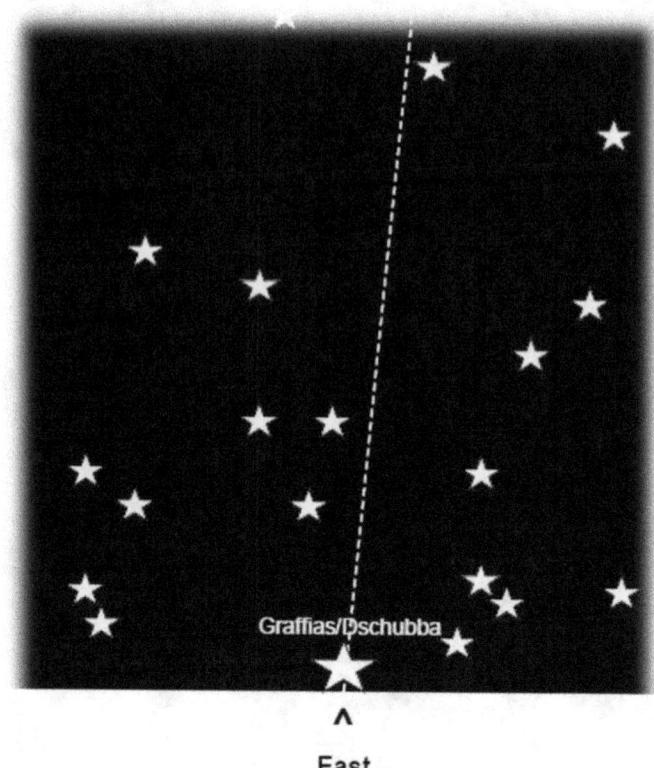

East

Eastern Horizon Giza – 1st January 2560 BC
(Stars brighter than magnitude 4)

Rising position of beta Scorpii (Graffias/Dschubba)

from www.fourmilab.ch

This binary star with, a magnitude brighter than 3 could then have provided a means for the surveyors to align this structure closest to the cardinal points of any Egyptian pyramid, which it is.

The Great Pyramid is also more closely aligned to due East than it is to due North.

The alignment of Khafre's pyramid indicates an exception to the use of this method, possibly due to the topography of the Giza site preventing a clear view to the east.

Legon's geometry of the Giza site which is based on Petrie's survey, shows how the position of the NE corner of Khafre's pyramid might have been based on a rectangle with a side of 250RC and diagonal of 500RC taken from the centre of the southern side of Khufu's pyramid.

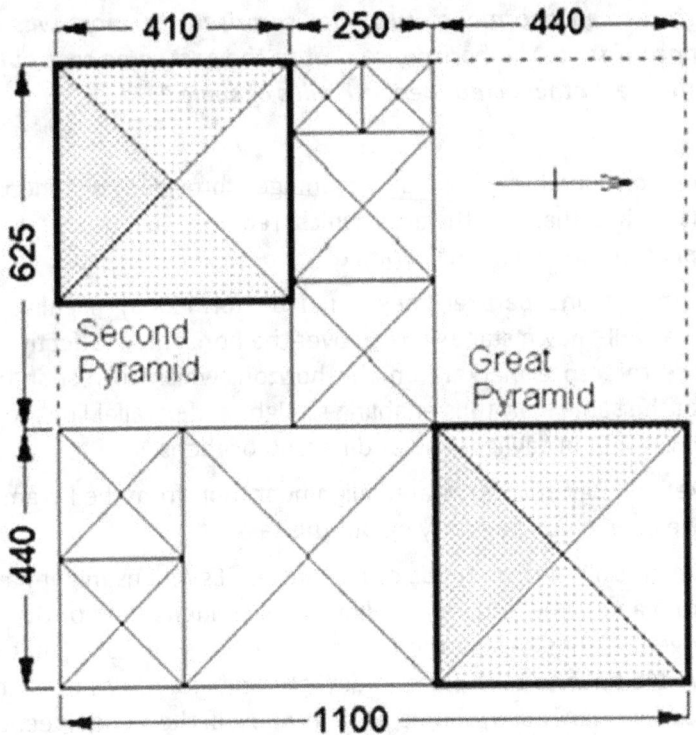

Modular Scheme Connecting the Second and Great Pyramids
Dimensions in Royal Egyptian Cubits

Thurston however notes that Khafre's pyramid has sides matching the setting position of Graffias/Dschubber and they are therefore not parallel with Khufu's southern side and he suggests that westwards might have been the alignment direction in this case.

Even if north had been an essential feature of pyramid positioning then it is more likely that a horizon star, at right angles to an approximation of north, would have been chosen as a more reliable and universally known reference to actually carry out the alignment. Horizon maps at the beginning of Sneferu's and Khufu's reigns, which include stars of magnitude 4 and below, clearly show beta Scorpii, aka Graffias/Dschubba, rising due east.

J. A. Michener in 'Caravans' describes the ease with which horizon stars can be seen in a desert environment.

"The stars seemed enormous, but what surprised me most was the fact that they dropped right to the horizon, so that to the east some rose out of the dunes while to the west others crept beneath piles of shale."

There are considerable practical advantages provided by choosing to align eastwards rather than northwards which requires the use of a complex and clumsy apparatus for each new pyramid.

Instead, straight and parallel lines could be formed by aligning two markers directly at a well-known star as it rose over the horizon. An effective option might be to place reference markers on the horizon, which correspond to the rising position of the same star, thus enabling straight and parallel lines to be formed in the same way, but in daylight and at different locations.

The Merket and Bay are two related alignment tools from the Pyramid Age, which would have been useful for carrying out this task.

The alignment skills and methods of the Ancient Egyptians might have developed initially from annual field marking following each inundation of the river Nile and for many generations before the Pyramid Age. Logic dictates that forming field squares would have to take place in daylight and when tens of kilometres apart, but formed simultaneously, should all be running in the same direction.

If this was the case, then the earliest alignments to beta Scorpii would only have approximated eastwards, with precession taking them closer to due east during the Pyramid Age, until providing an almost exact and unique reference when the Great Pyramid itself was constructed.

If any method of alignment sought to provide a means for ensuring that pyramids had accurate square bases directed to cardinal points, then it is reasonable to assume that the Ancient Egyptians would have chosen the most practical and effective method of achieving this aim.

The Virtual Apex Method for Accurate Pyramid Construction

The provision of a reference for accurately forming the sides of a pyramid without the apex being in place can be achieved by exploiting the geometry of a symmetrical pyramid.

An accurate pyramid has the same proportions throughout its height, in either its profile or diagonal cross section.

The centrepoint-corner cross section of a pyramid forms a vertical right triangle, with the hypotenuse forming the corner edge. This triangle is either a right isosceles triangle, when the base centrepoint-corner dimension equals the height, or a right triangle with a base plus or minus a fraction of the height.

This relationship can be described as the height : centrepoint-corner ratio.

In a symmetrical pyramid, the height : centrepoint-corner ratio found at the base, is repeated at any height above the base. This consistent geometry, if systematically applied, can provide the corner positions of a pyramid at selective heights and from these the corner edges.

For example, suppose we want to make a solid symmetrical pyramid 100 high, with base corners 110 from the centrepoint and without using the apex as a reference. The height:centrepoint-corner ratio in this pyramid is 10:11, as the corner is the height + one tenth of the height from the centrepoint

$$(100 \div 10 \times 11 = 110)$$

At any height above the base the pyramid corner position must maintain this ratio if the pyramid is to remain symmetrical. Therefore at a height of 10 for example, as the height **remaining** to the apex is 90 the outer corners will be 99 from the centrepoint. (90÷10 X 11 = 99) If these upper corners are positioned and then connected by a straight line to the corners of the base square below, the pyramid corner edges will be accurately defined to this height. Straight lines running between the corner edges will automatically define the pyramid faces at any height.

The same procedure can be repeated at heights of 20, 30, 40 and so on, as each leaves a **height remaining to the virtual apex** exactly divisible by 10. The corner positions at each of these heights, when connected to the corners below, exactly define the pyramid corner edges and from these the pyramid faces.

Each of these edges, as they are extended upwards will intersect at the **virtual apex** exactly 100 above the base, with each flat face inclined in this example at an angle of **52°7'33"**

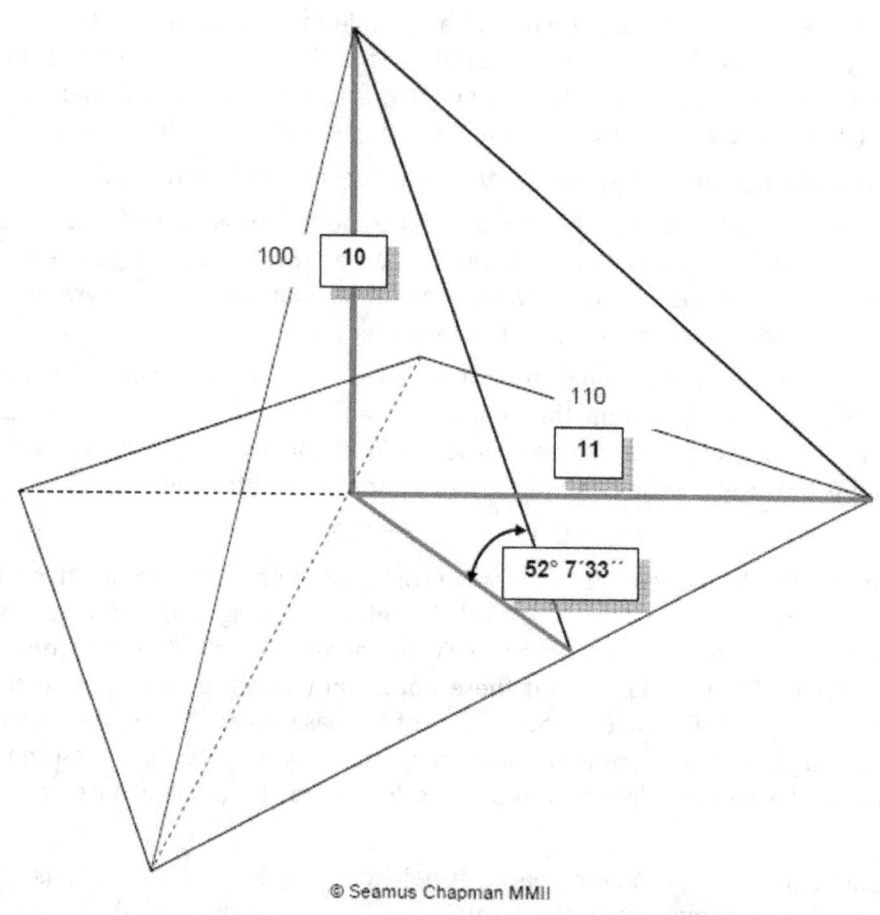

Relationship between height:centrepoint-corner ratio and a pyramid slope angle

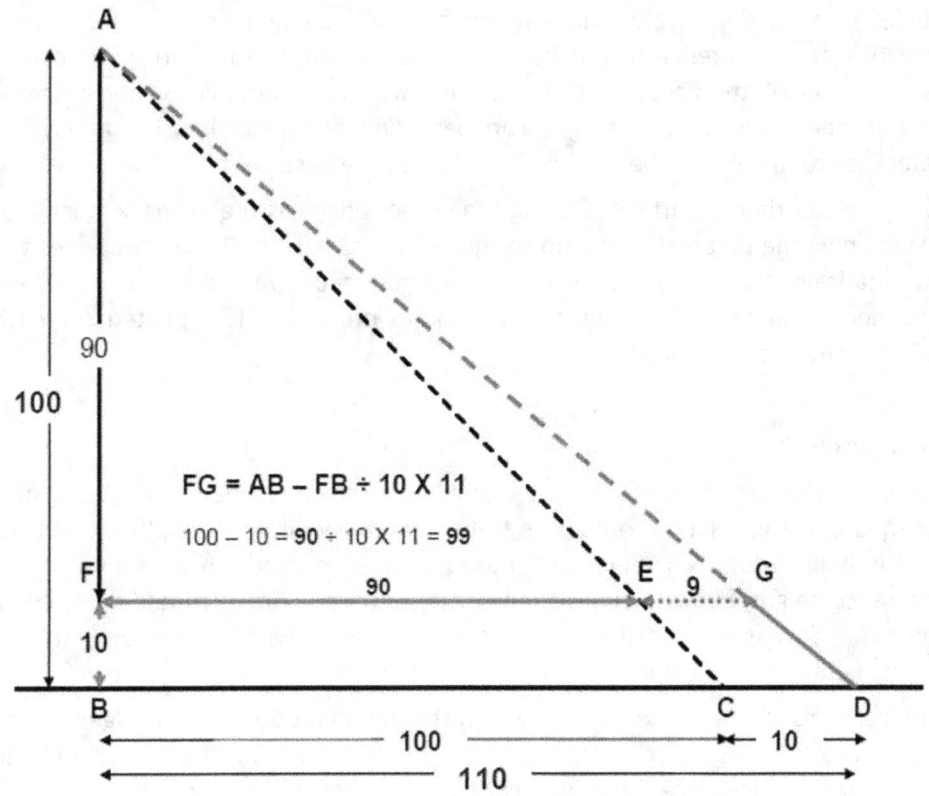

Diagonal Cross-section

Geometry of Symmetrical Pyramids

Pyramids of any shape or dimensions can be formed using the virtual apex method, by employing different height:centrepoint-corner ratios in the same way.

However if the chosen dimensions for a pyramid create a height:centrepoint-corner ratio based on whole numbers, both the reference heights and corner edge positions can be calculated using only basic arithmetic.

When a method is known for providing an exact corner edge without the apex in place as a reference, an accurate pyramid can be formed using this as the primary guide. In a solid pyramid, this can be achieved by first constructing a core structure to a reference height based on the chosen height:centrepoint-corner ratio, which will then act as a platform from which the pyramid corner positions and corner edges to this height can be found and placed, using a physical reference running from the platform, straight to the base corner.

A façade can then be attached using the corner edges as the primary means for maintaining the external shape up to the reference height. The corner edges will automatically form the four sloping sides when a straight line is stretched between them at various heights. The same procedure is repeated for each subsequent reference height.

Implications

The core structure can be constructed to a known reference height, requiring only its squareness in relation to its base to be maintained and the height monitored. If it is built around an un-cleared base, the centrepoint can be found at the reference height, by taking lines from corner to corner. These should be of equal length and cross the centrepoint at right angles, with an opportunity for adjustments to be made if required.

The core provides a platform from which the pyramid outer corner edges can be defined to this height. The marking of the face edge on the top of a completed course takes place once the whole course is in place and the corner edges have been cut on the corner blocks.

A stringline running between the accurately cut corner edges describes the face edge position exactly at each course height.

Partially prepared façade casing blocks can be placed routinely on a subsequent course, using as a guide the face edge position clearly marked as a straight line on the top of the course below.

Casing blocks preparation therefore only requires each to have a level base, sides which fit to their neighbours and be of sufficient depth to maintain the face slope.

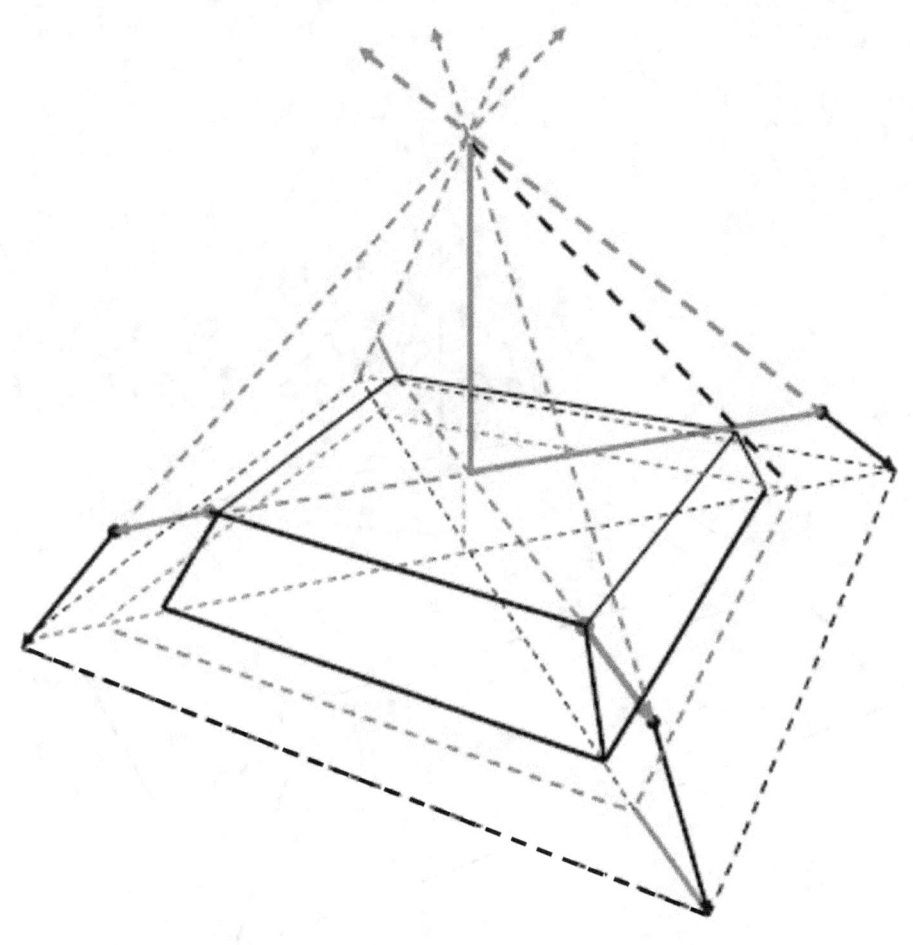

A core structure provides a platform for placing markers to the pyramid outer corners

The height: centrepoint-corner ratio is used to find the dimension to the pyramid outer corners at each reference height

A perspective view of a solid pyramid
Reference Core completed to a Reference Height

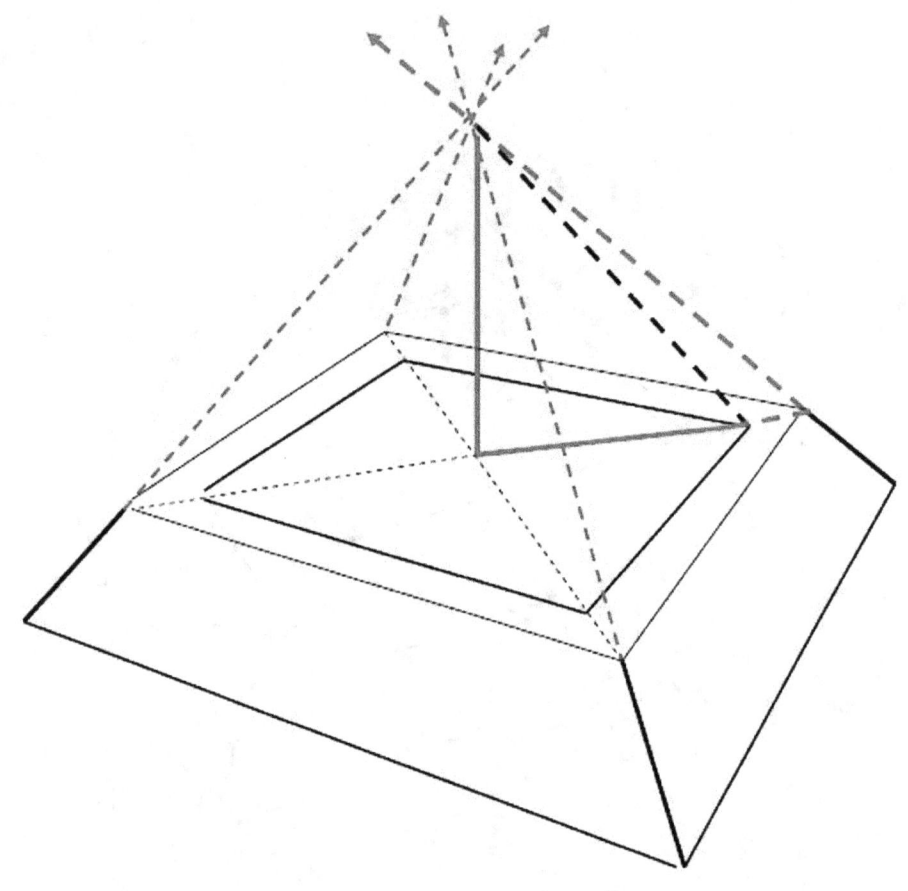

**A pyramid with the façade added to the
height of the reference core using the corner edges as
the primary reference in controlling the shape**

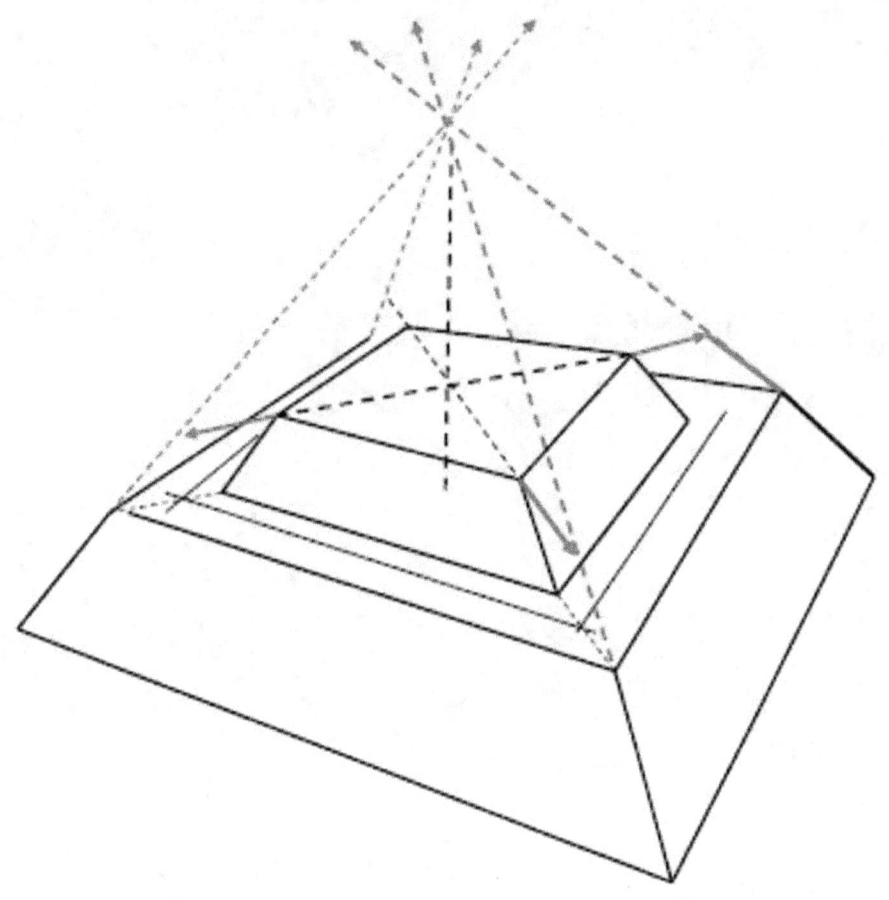

A pyramid with the core taken to a new reference height

Reference Squares and the Virtual Centrepoint

Although the virtual apex method does not use the base centrepoint for controlling the pyramid shape, it is vital that the corners of the base square are a known distance from the **virtual centrepoint**.

All squares have a consistent relationship between their side and diagonal dimensions, as side length X √2 = diagonal length, side length divided ÷ √2 = centrepoint to corner and diagonal length ÷ √2 = side length. However, some squares also have a side-diagonal relationship where each dimension is close to whole numbers.

For example a square of side 70 has a diagonal of 99 (70 X √2 = 98.99.), a square of side 99, a diagonal of 140 (99 X √2 =140.007) and so on.

Application of this relationship can provide some of the information about squares necessary for pyramid building, but without a need for an understanding or awareness of √2.

For example, the exact diagonal length of a square with sides of 408 is:

576.99. (408 X √2)

However 408 is also (70 X 3) + (99 X 2), which converts simply to

(99 X 3) + (140 X 2) = 577

In the same way, a square with sides of 495, which is also 99 X 5, converts to a diagonal 140 X 5 = 700 and each of these examples is reversible.

By combining these results, other side or diagonal dimensions can be found, including those not directly divisible by either 70 or 99.

For example a square with a side of 87 has a diagonal of 123 as:

495-408 = 87 and 700 – 577 = 123

An alternative application suggested by Legon allows the diagonal or side length of certain squares to be determined, by multiplying one dimension by 99 or 70 and then dividing by 70 or 99, respectively.

For example the diagonal dimension of a square with sides of 67 cubits using √2 and 99/70 as a substitute for √2

$$67 \times 99 = 6633 \div 70 = 94.757 \text{ using } 99/70$$

$$94.752 \text{ using } \sqrt{2}$$

And 94 cubits, 7 palms + 3 fingers using long division shown below

67X99=6633   70) 6633 (**94 cubits r.53** or (94.7571)

$$\underline{630}$$

$$333$$

$$\underline{280}$$

r.53 = 7 palms + 3 fingers

However as the side length of a square with a known dimension from its centrepoint was possibly of more interest to the Ancient Egyptians, 99/70 can be employed once again here as a substitute for √2.

The centrepoint to corner dimension is therefore:

$$67 \times 70 = 4690 \div 99 = 47.37$$

99)4690(47 cubits r.37

$$\underline{396}$$

$$\underline{730}$$

37 = 5 palms, 1 finger

Other examples

117=165 r.33 = 165 cubits, 3palms, 1+3/4 fingers (7/4) 165.47 or 165.46 (x√2)
326=461 r.4 = 461cubits, 0 palms, 1+3/4 fingers (7/4)461.057 or 441.03 (x√2)
176=248 r.64 = 248cubits, 6 palms, 1+3/4 fingers (7/4)248.91 or 248.9 (x√2)

Calculating the side length of a square from the diagonal dimension using 70/99 is not as elegant, but is unlikely to have been a necessity in pyramid building.

Pyramid profiles and diagonal cross sections

These whole-number side to diagonal relationships also provide a simple method for converting the base dimensions of a specific pyramid profile into a known centrepoint-corner dimension of the same pyramid. This information can then be used to calculate simple and consistent height : centrepoint-corner ratios.

For example a pyramid with a profile of height 100 and base 198, has a diagonal cross section of height 100, a diagonal of 280 and a centrepoint-corner dimension of 140.

Side 198 = 99X2 therefore centrepoint - corner = 70X2 = 140

This can be expressed as a height:centrepoint-corner ratio of 100:140 or 10:14.

As the height to centrepoint-corner proportions are now known a pyramid can be built using the virtual apex method to accurately form the shape, with this example having a profile angle of **45° 17′17″**

Implications

Application of a simple formula provides a means to determine the diagonal dimensions of squares and enables pyramid shapes to be based on either their profile or diagonal cross section proportions and provides a means to build them accurately, using the virtual apex method.

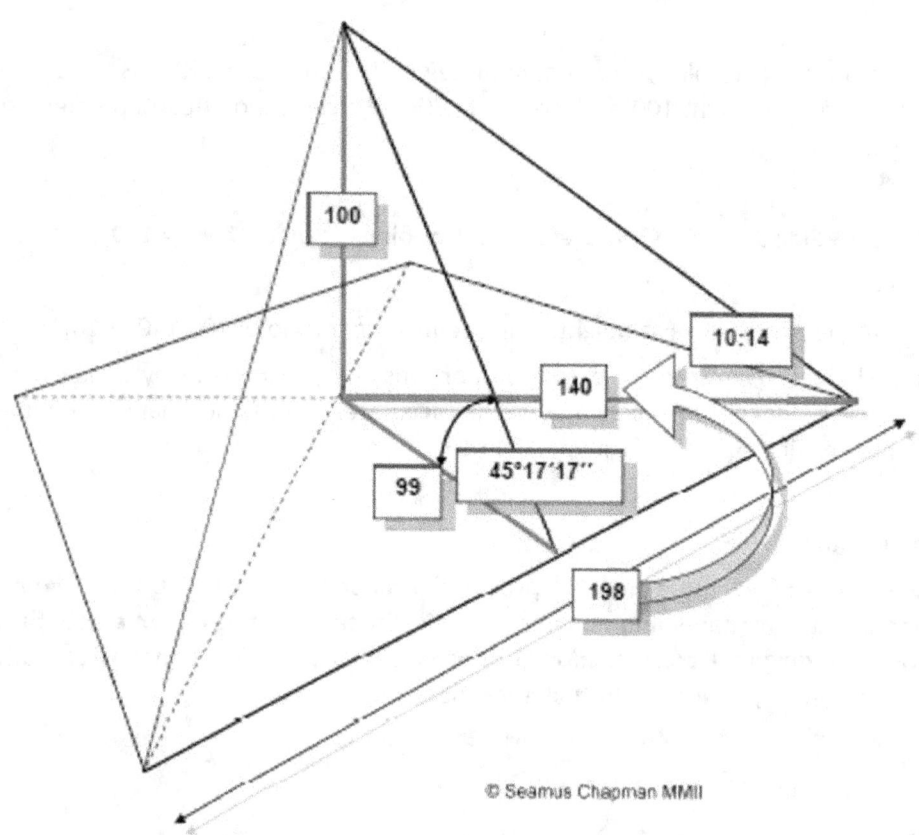

**Converting a side length to a centrepoint-corner
dimension using whole numbers**

Forming Right Angles and Diagonals
of known Dimension from a Virtual Centrepoint

Another application of this knowledge follows almost automatically. For example, on a straight line, peg 1 and peg 2 are placed 70 apart. Lines of length 70 and 99 are attached to each peg respectively with peg 3 positioned at the point where the two lines meet. The lines are reversed on each of peg 1 and 2 and peg 4 is found in the same way as peg 3.

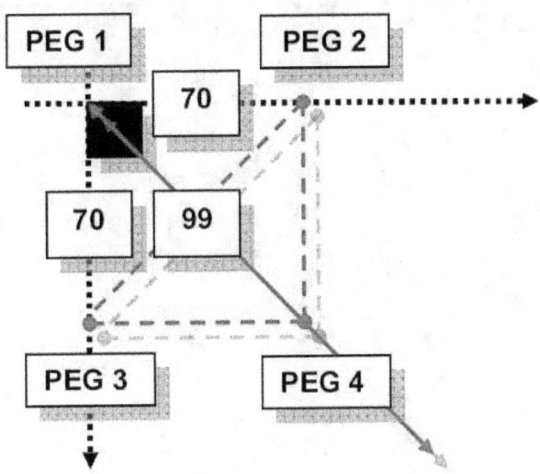

A right angle and diagonal is formed at peg 1.

If this action is repeated on lines of equal length which extend at right angles from Peg 1, any larger square can be accurately marked out with partial diagonals formed without having to cross the centrepoint.

If the lengths of the sides of the chosen square also fit the side-diagonal formula described above, each of the corners will have a partial diagonal which is a known dimension from a virtual centrepoint.

It would also be possible to place the corner positions of any square with a required dimension from a virtual centrepoint simply by adding to, or subtracting from the partial diagonals of a primary reference square which has a known dimension to the centrepoint.

For example to form a square with corners say 160 from a virtual centrepoint, a reference square with sides of 280 (70X4) is formed first, with partial diagonals known to be 198 (99X4÷2) from the virtual centrepoint.

Markers placed 38 inside these partial diagonals provide the corners of a square the required dimension of 160 from the centrepoint, with side lengths an **unmeasured** 226.27.

1. Forming a right angle and partial diagonal of known length
2. Using a reference square to form any square with a known dimension from its corners to the virtual centrepoint

X = chosen dimension deducted

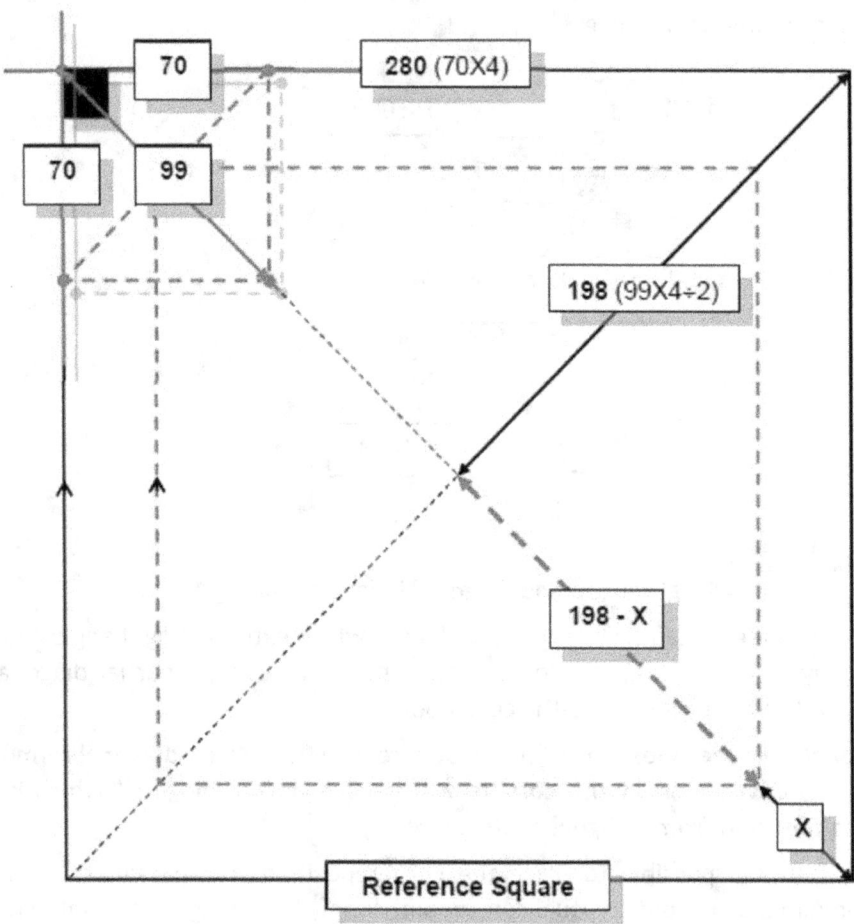

Implications

Preparation of a site on which a pyramid is to be constructed, need only be in those areas necessary for surveying and marking the corners of a reference square with a known dimension to the virtual centrepoint.

Discovery in Ancient Egypt

There is no evidence to suggest that the Ancient Egyptians were aware of the roots of numbers. The relationship of side 70 to diagonal 99 or side 99 to diagonal 140, might have been discovered as a result of preparing and measuring large squares. The larger the square formed, the more obvious and accurate the result.

The Egyptian unit of area the Aurora, is defined as a square with sides of 100 Royal Cubits (1 RC = 0.524m). It was a standard used for generations, most probably to subdivide the Nile floodplain accurately after each annual inundation.

If the Ancient Egyptians were routinely marking squares on this large scale and using this measure, it is quite possible they discovered empirically, either that a square with sides of 99RC has a diagonal of 140RC, or a square of side 70RC has a diagonal of 99RC.

Legon suggests that evidence for this discovery might be found in and explain their choice of 7 fractions for the Royal Cubit which gives a square with sides of 10 Royal Cubits or 70 palms a diagonal of 99 palms.

This square would be ideal for forming accurate right angles for large squares and provides a basis for the side lengths of reference squares having known dimensions to the centrepoint.

The Ancient Egyptians appear to have been aware of this relationship on a smaller scale as a square of side 10 was assigned a diagonal dimension of 14. It is reasonable to assume that this might have been a result refined from their knowledge of a larger square with sides of 10RC, with 14 representing a rounding down of the diagonal from 99 palms to 98 palms and then dividing 98 by 7.

This relationship might also have been the basis for discovering a method of calculating the circumference of a circle from the radius dimension without understanding Pi, evidence of which can be found in the proportions of the Great Pyramid. The relationship between the centrepoint-corner and height in the GP is 10:9 and the diagonal 20:9. The relationship between the height and the side is therefore:

$$\frac{20}{9} \times \frac{99}{70} = \frac{22}{7}$$

The relationship between the radius to the circumference of any circle is 2Pi and an Ancient Egyptian equivalent might therefore be:

$$99/70 \times 40/9 = \frac{44}{7}$$

Circles with radius 28 and 126 have circumferences of 176 and 792 as:

Radius 28 X 44 =1232÷7 = 176

Radius 126 X 44 = 5544÷7 = 792

Radii with a product not divisible by 7 will provide a whole number of RC followed by a whole number of palms based on any remainder when using long division. For example:

Radius 23 = 144, r4 or 144RC + 4 palms circumference

Radius 117 = 735, r3 or 735RC + 3 palms circumference

The drawing below shows the relationship between diagonal and side dimensions of squares and the circumference of circles which have the same radius based on the formula 99/70.

If these relationships were known to the Ancient Egyptians it might explain their choice of dimensions for the Great Pyramid.

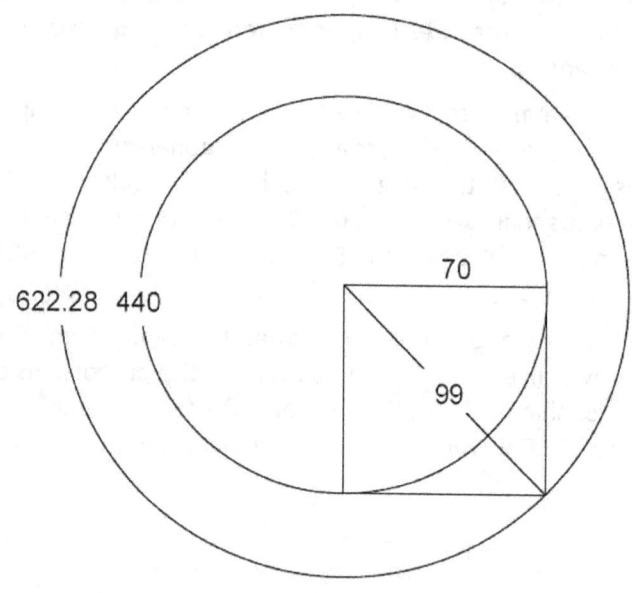

70 X 4 =280 = Height of Great Pyramid

70RC X 44/7 = 440RC = Base side length of Great Pyramid

99 X 44/7 = 622 r2 = 622RC+2palms = Diagonal of Great Pyramid base.

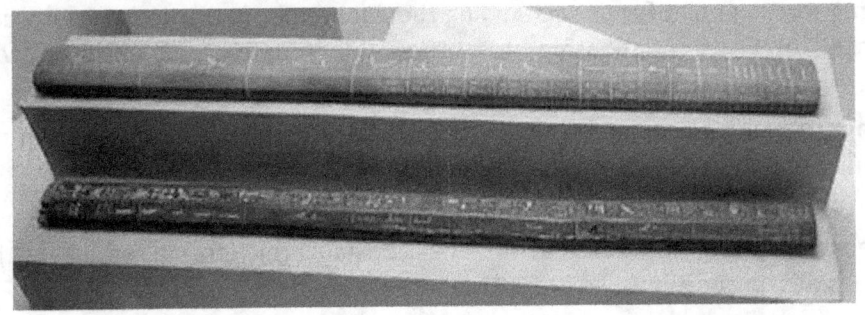

Contemporary Royal Cubit divided into 7 palms of 4 fingers

The Aurora

100 Royal Cubits = 700 palms

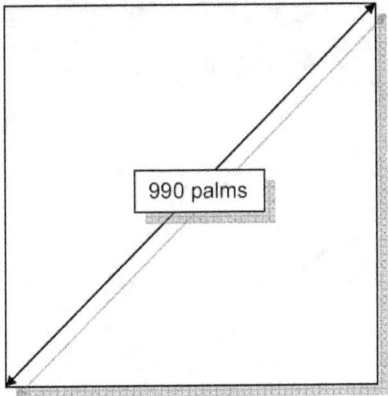

990 palms

10 Royal Cubits = 70 palms

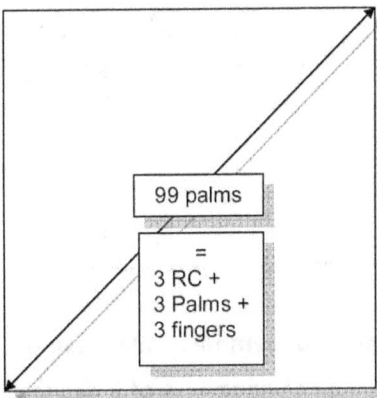

99 palms
=
3 RC +
3 Palms +
3 fingers

Constructing squares with related side and diagonal dimensions might also have been a starting point for understanding the geometry of a vertical right isosceles triangle and its relationship to others of different dimensions, when superimposed to the same apex.

In this arrangement it can be clearly seen that - the difference in height of one, when added to its base dimension - equals the base dimension of the other.

By superimposing other right triangles of equal height in the same way, observations can be made about the constant relationship between their dimensions, proportions and areas at different heights.

Recognising this constant provides enough information from which a method of forming the hypotenuse of a right triangle, from its base to a virtual apex at a specific height might be discovered.

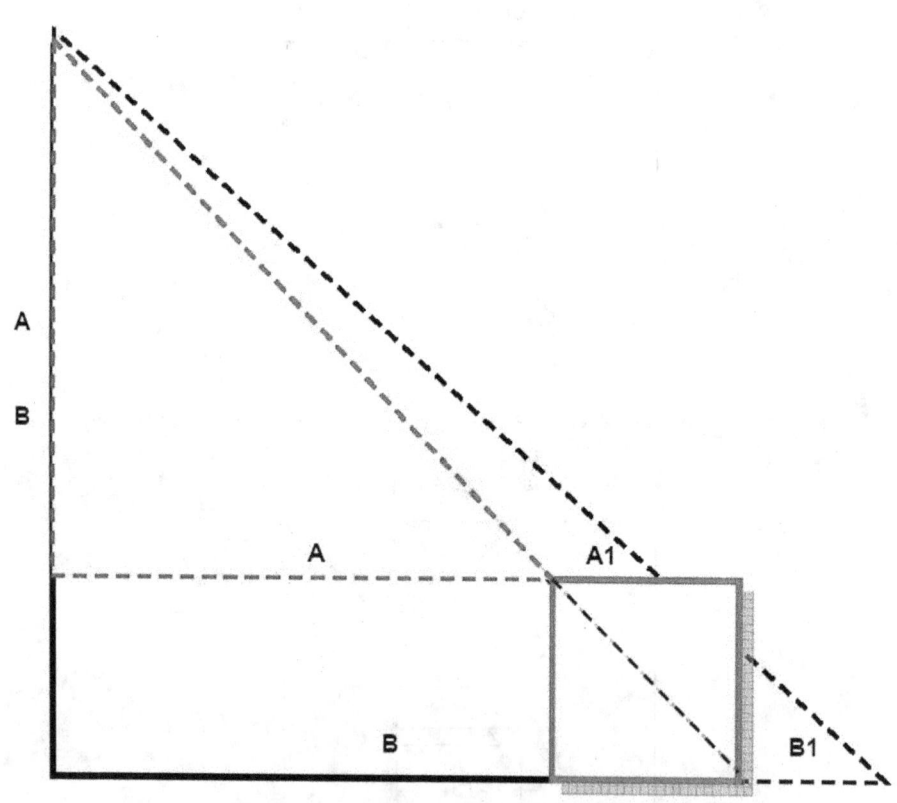

Geometry of Symmetrical Pyramids
Understanding the geometry of a right triangle

A systematic process for constructing a solid stone pyramid accurately

1. Position the corners and partial diagonals of a suitably sized reference square with a known dimension to the centrepoint.
2. Fix the Pyramid base corners on the partial diagonals the correct dimension from the centrepoint. Mark the pyramid base line.
3. Build a solid core within the base square to a known reference height.
4. Confirm the centrepoint on the reference core top and mark the pyramid outer corners on extended diagonals at this height.
5. Connect the upper corner positions to the base corners with a straight line to provide an exact corner edge reference aligned to the virtual apex.
6. Lay the first course of façade blocks with the casing aligned to the base square and cut their top surface level and flat.
7. Cut the corner edge on the casing corner blocks to fit the reference line.
8. Mark the top of the casing course with the face edge position, based on a straight line running between the pyramid corner edges at this height.
9. Lay the following course with the base of the casing blocks aligned to the face edge line marked on the top of the course below.
10. From the corner edges, mark the face edge position on the flattened top of the whole course.
11. Add further courses of the façade in the same way up to the height of the reference core.
12. Construct the core to a new reference height and repeat the process of forming the corner edges and adding the façade.
13. Core and façade construction continues to a height at which the pyramid corner edges can be readily extended to the apex with straightedges.
14. From the capstone down, cut the pyramid faces to fit straightedges running between the face edges marked on the top of each course of casing blocks.

A video demonstrating the application of the construction sequence described above can be seen on **www.youtube.com** as **'The Great Pyramid Building Challenge'**

Despite its small size, this is the first and only pyramid constructed since the Pyramid Age, with the same conditions and using only the tools available to the Ancient Egyptians. No one has attempted since to duplicate it.

Access Ramps and Platforms for the Delivery and Installation of Materials for the Core and Façade of Pyramids

The 'Golden Steps' of the Chilkoot Pass - 1890's Yukon Gold-Rush

Over 30000 'Stampeders' (from more than 100000 who began) completed the 1000 km land, sea and river journey to the Yukon which included crossing the 300 metre high Chilkoot Pass with a minimum of 1 ton of supplies. It took 40 trips up the 1500 ice steps with a 25 kg load and 39 trips sliding back down for the next; an overall effort equalling that of Pyramid building.

A perpendicular ramp attached to one face can be used to deliver materials efficiently only during the earliest stages of pyramid construction. It would have been ideal for construction of the core to its first reference height and also allow any massive blocks to be taken to this height. However the delivery and placing of materials for the façade was a critical feature in pyramid construction and the ramp used has to achieve a number of objectives. It would be used to not only deliver the casing and backing stones to every height including the apex, but must also provide external access to all the pyramid faces at each course level. The final fitting of casing blocks was achieved by pushing them into place from the front and without lifting. This ramp would also be used to provide access for core construction materials at higher levels, after each section of the casing was completed to a reference height.

A pyramid ramp can only be designed in one way to meet the objectives of low gradient, external access to all faces at each course level and continue in the same way to the apex.

A platform, level with the top of a completed course, extends around all four faces of the pyramid. Its base rests on the surplus stone of the casing courses below which extend as steps outside the pyramid face edge and it is fed initially from a single perpendicular ramp. When the volume of material to maintain the perpendicular ramp becomes excessive, a new ramp is deployed which runs parallel to the outer edge of the platform. It spirals up and round the pyramid as new platforms are added as each façade course is completed. If the wall slope of both the platform and ramp is carefully controlled as each rise to the next course, the geometry ensures that after one circuit the access ramp base will sit exactly on top of what had been the platform of the previous circuit.

Instead of running out of space, the ramp begins each new circuit in exactly the same way as the one before.

This unique design of platform and spiral ramp exploits the pyramid shape, can be of different widths and ramp gradient and continues in the same way to the apex. The design is safe and economical with materials, with the ramp rising throughout at gradients of less than 5%.

For example a ramp of this design, for a pyramid with a base 230m square and 150m tall, would make 6.5 circuits of the pyramid, if connected to a perpendicular ramp which completing the first 20m and be a total length of 3000 metres to the apex. If the width of the spiral ramp and platforms were each a generous 2m, it would require less than 250,000 cubic metres of material to construct it

Compare this with a single perpendicular ramp of the same length, with the same gradient and to the same height, having access at only one point on one pyramid face and requiring almost 1,500,000 cubic metres of material to build it.

The length of the external spiral ramp has two positive effects. As a conveyor it would create order and space for the block moving teams and its pulley effect would reduce the effort to move a block upwards, by increasing the distance travelled horizontally. Turning corners would have been the most difficult aspect of moving blocks and this would have been assisted by taking pulling ropes beyond the corner and using rams to push the largest blocks from behind. It is likely that specialist teams would be strategically placed at each ramp corner to ensure the flow rate was maintained. Blocks for the core could be taken to any point either along the platform or directly onto the core itself, but more significantly the design would allow block moving teams, when arriving at the top of the ramp with façade blocks, to be directed either left or right along the level platform to their appropriate position on the pyramid. Installation teams could then take the backing or casing blocks and speedily slide them into place.

Because the platforms were mounted to the whole exterior at each course height, casing blocks at each level could be placed last with an additional feature enabling them to be taken first to the point furthest from where the access ramp connected to the platform. Immediately a casing block was in place, the platform could be raised to the top of the course with the top of each installed casing block levelled, making it ready for marking with the face edge position when the whole course was complete. Even the course of the final side could be progressively levelled and the platform raised, as the last blocks were being brought into position.

These actions would prevent any delay in block delivery or installation of the next course, as time would only be needed to extend the access ramp to the new platform height with this work carried out as the final section of blocks were being levelled and marked. The volume of material required to maintain the ramp and platforms can be delivered at the same time as material for the pyramid itself. Both platforms and ramp can be of any chosen width at particular heights, but sufficiently broad to allow casing blocks to be pushed into position horizontally and workers to pass others freely. For example as construction approaches the upper areas, the widths could be reduced to 1 metre each without compromising either block delivery or safety. Even today workers routinely work on construction projects at great heights on platforms only 30 cms deep.

The comprehensive access provided by the platform and ramp system described, ensures that the masses of blocks for the core can be delivered rapidly and efficiently and the delivery and fitting of the façade can be carried out routinely and accurately.

It can be noted that a delivery system and access platforms which completely cover the façade of a pyramid is only possible when the alignment method does not require the completed lower sections to be visible.

Remains of a brick platform/ramp at Karnak

Evidence of platform/ramp systems completely cloaking a structure can be found at the Karnak Palace complex at Luxor and from contemporary accounts of the Taj Mahal at Agra which was completely covered in a brick scaffold during its 20 years of conversion from a Hindu Palace to a Mughal Mausoleum.

A Ramp and Platform System for delivering materials to Egyptian Pyramids

In any plan view taken at any height, the platform, the point at which the ramp connects to the platform and the ramp itself, maintains the same features

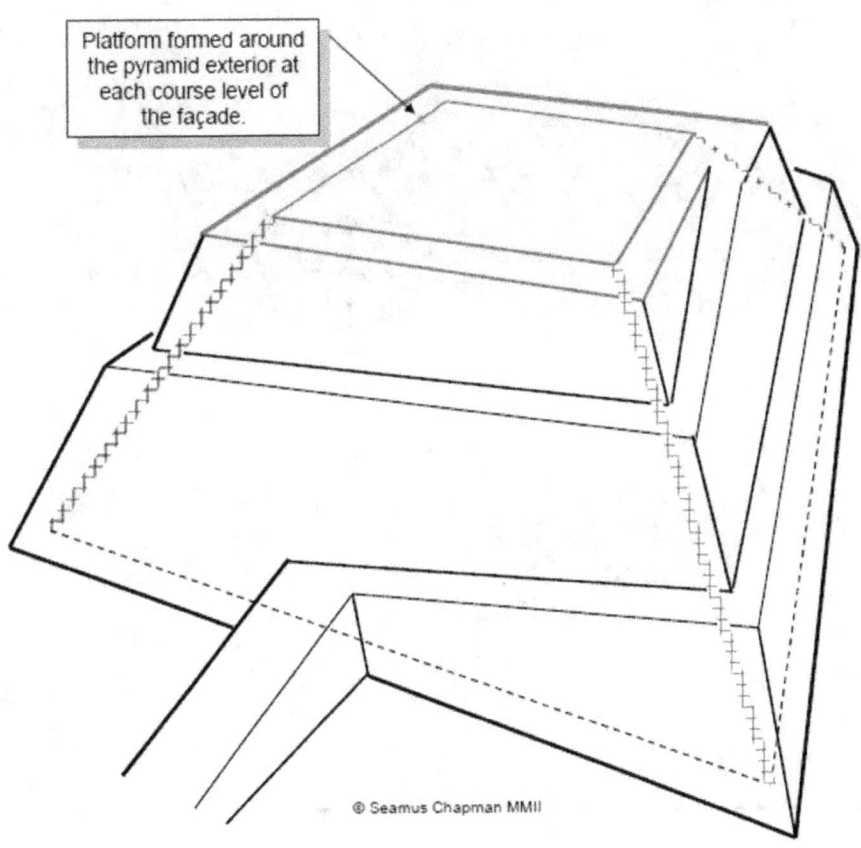

Platform formed around the pyramid exterior at each course level of the façade.

An integrated Platform and Ramp System for the delivery and fitting of construction materials to Pyramids in Ancient Egypt

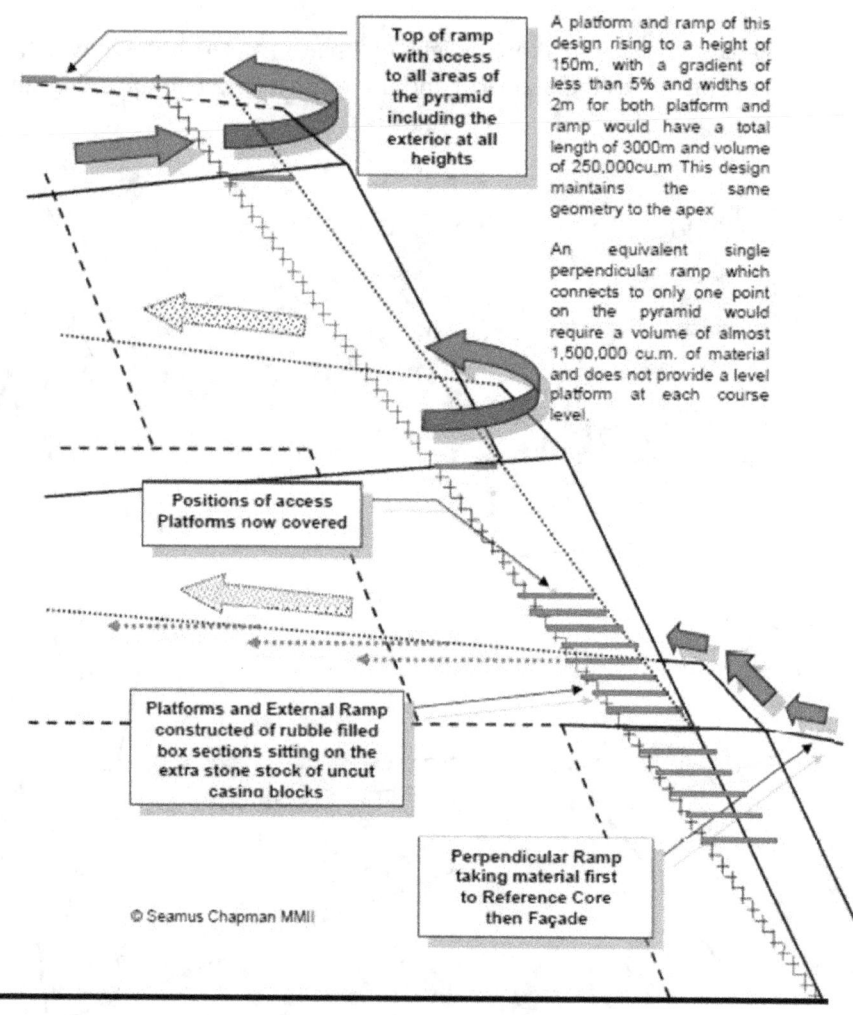

A pyramid ramp cross-section showing platforms at each course level during construction

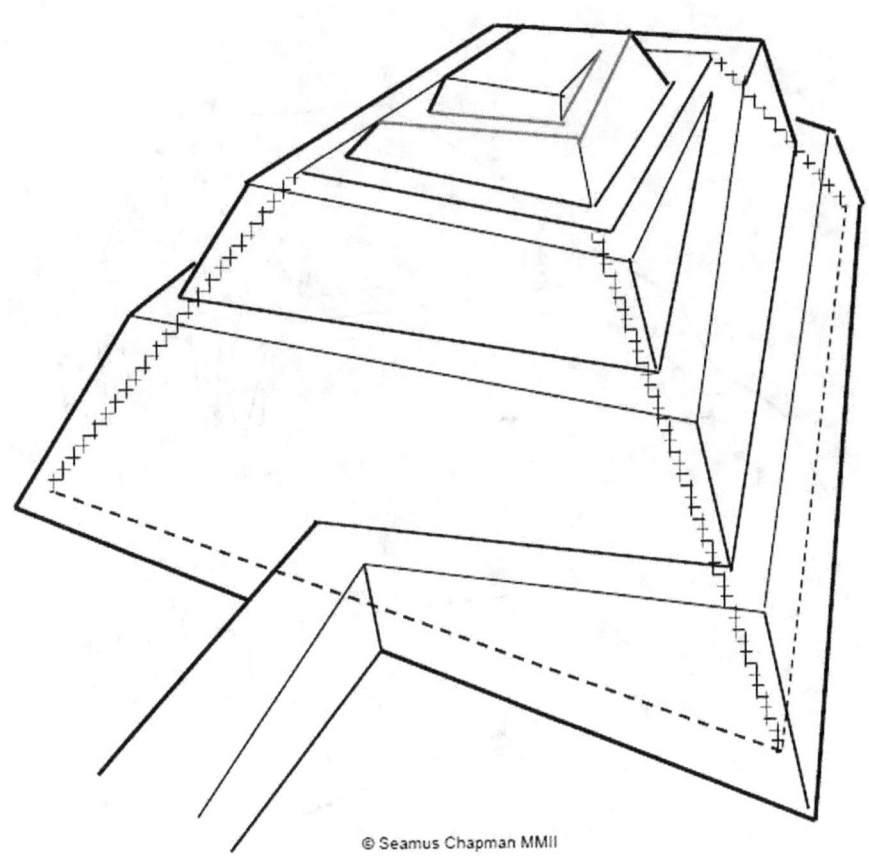

**A pyramid ramp and platform used
to take material to the reference core**

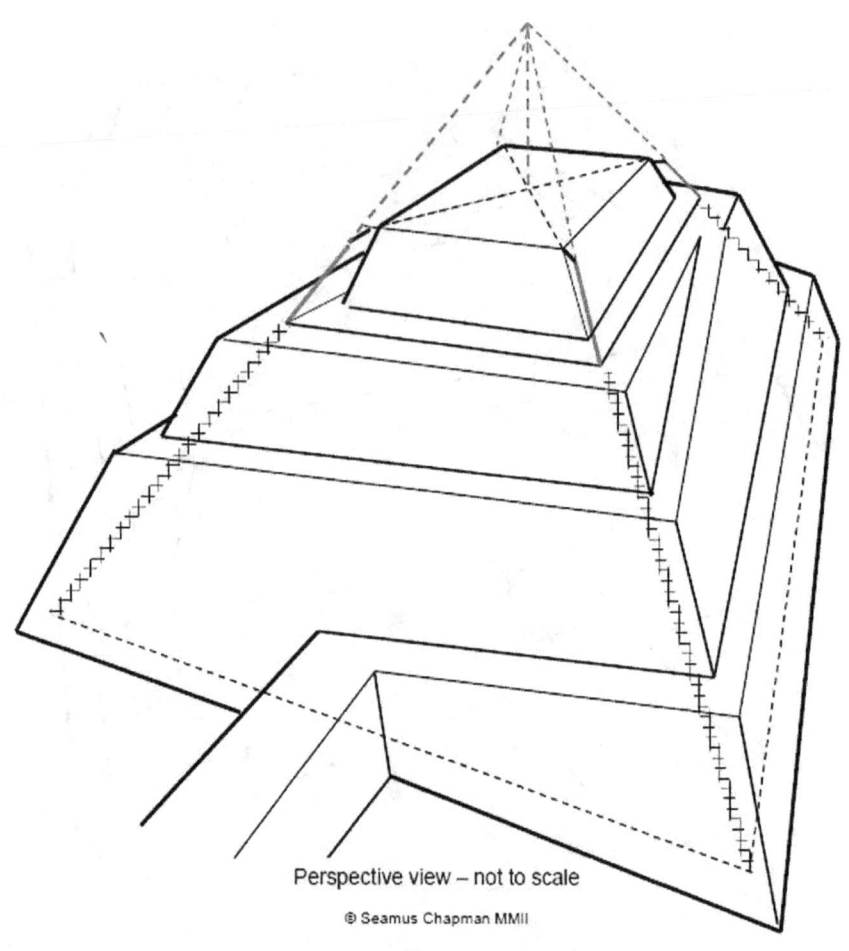

Perspective view – not to scale

© Seamus Chapman MMII

Determining the corner edge position at a new reference height

An integrated Platform and Ramp System maintaining the same geometry to the apex of a pyramid

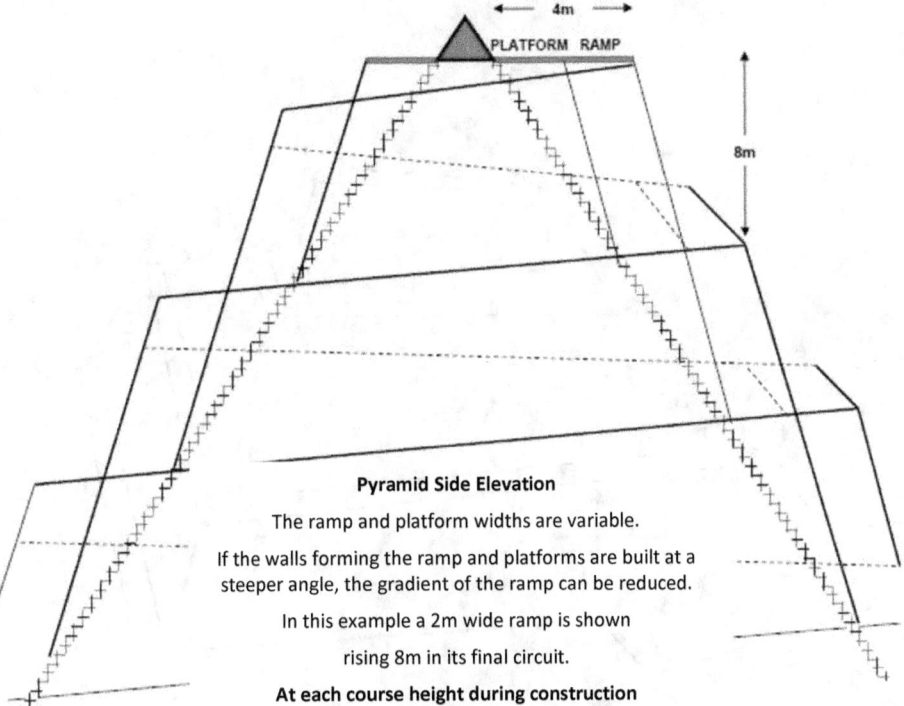

Pyramid Side Elevation

The ramp and platform widths are variable.

If the walls forming the ramp and platforms are built at a steeper angle, the gradient of the ramp can be reduced.

In this example a 2m wide ramp is shown

rising 8m in its final circuit.

At each course height during construction

a level platform runs around the pyramid

On completion of the pyramid, the ramp and platforms provide comprehensive access as they are dismantled to reveal the casing stones, which are cut back to the face edge positions marked on them during construction.

Not to scale

© Seamus Chapman MMII

Fitting the Capstone
Pyramid Platform and ramp at the Apex

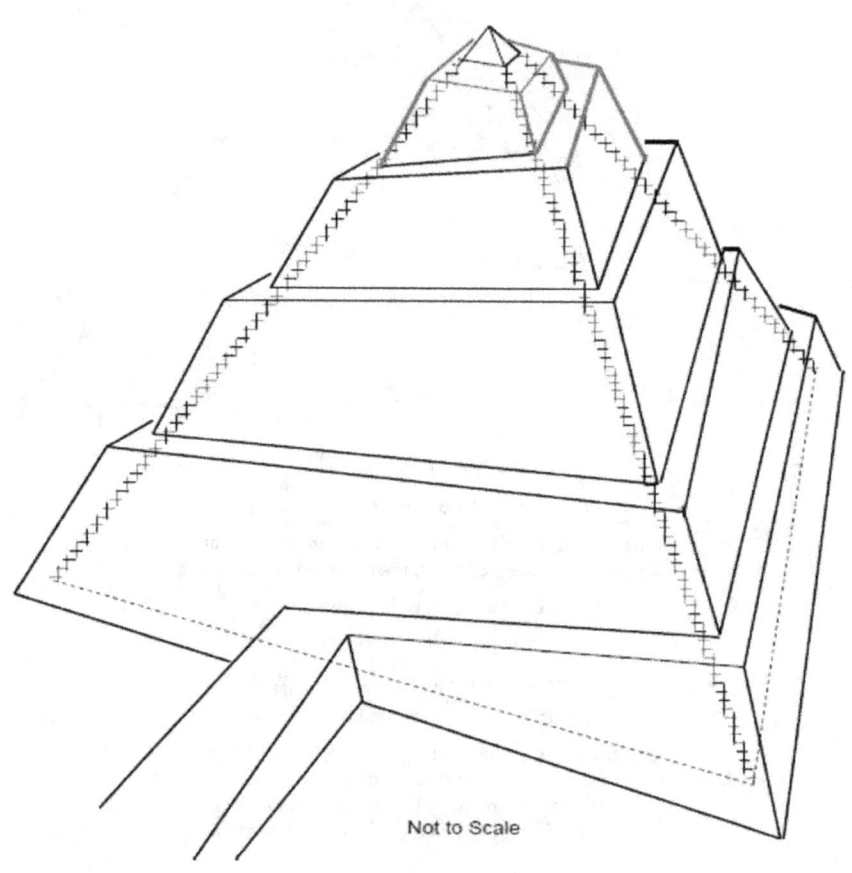

Not to Scale

Preliminary Conclusions

It has been shown that a fundamental of accurate pyramid construction is how the corner positions at different heights are vital for forming the shape of any pyramid.

The virtual apex method provides a means for determining these at reference heights, without requiring the apex for reference.

The side to diagonal relationship of 70 to 99 provides a method for readily calculating the centrepoint-corner dimension in some squares using basic arithmetic. Reference squares based on this relationship provide a means to form any square with corners a known dimension from its virtual centrepoint.

If a pyramid shape is to be based on a specific profile, the chosen dimensions must first be converted to a diagonal cross section in order to provide a means to build it using the Virtual Apex method. If the relationships of a 70-99 Reference Square were the only means to make this calculation, adjustments would have to be made in some cases, to the centrepoint-corner dimension, to ensure that the virtual apex ratio is based on a simple fraction. Any adjustments will as a consequence, automatically change the profile. The height:centrepoint–corner ratios of Egyptian pyramids, might therefore provide the answer to why some profiles are found, which do not describe an exact seked ratio.

These 'errors' might not be errors of measurement, but a result of these adjustments, which provide the construction processes necessary to build them accurately and systematically. It is also true that the majority of markings found on partially exposed pyramids, are either height or diagonal references.

In the absence of a viable alternative, this suggests that the Ancient Egyptians were aware of the features of Reference Squares and their Virtual Centrepoints, together with the Virtual Apex method for building accurate symmetrical pyramids and applied them successfully throughout the Pyramid Age.

The delivery of materials to all parts of the pyramid including the exterior can be achieved when the access ramp is attached to platforms, which run round the whole pyramid at each course level.

The integrated platform and ramp system described provides this comprehensive access, to all parts of a pyramid and at all heights, including the apex and is less than one tenth of the volume of an equivalent perpendicular ramp.

Examples of Egyptian Pyramids

The pyramids are presented in their accepted order of construction.

The exceptions are the Stepped pyramid, which is not a true pyramid, but was the first large stone monument constructed in the Pyramid Age.

The Meidum pyramid, which followed was also built as a stepped structure in its first phase, but had later additions which enlarged the first phase and included a smooth pyramid façade. This structure is included in the date order of when the second phase was thought to have taken place.

The two largest and most well-known pyramids, Khufu's and Khafre's are examined in greater detail, including each of the steps, which might have been employed to determine their height : centrepoint-corner ratios.

It must be considered however that the vital dimensions necessary to apply the virtual apex construction process, were decided from the outset in every case and the profile dimensions are only an automatic consequence these choices.

1 Royal Cubit = 53.4 centimetres = 20.6 inches

The Southern 'Bent' Pyramid

This pyramid is unusual in that it changes shape part way to the apex. Whether this change was part of an original plan is unknown, but where the change in shape takes place provides valuable information.

It has base side lengths close to 360RC, a measured face slope averaging 54º44' for the lower section and corner edges therefore rising at a height:centrepoint-corner ratio of 1:1, or 45º.

This is the simplest of ratios to apply using the virtual apex method, as the centrepoint-corner dimension decreases by an amount equal to the increase in height. This unique corner edge ratio removes the need for a reference core to be built to specific heights to define the corner edge as this can be determined from the core at any height and permits the core and façade to be placed at the same time.

If construction had continued with this ratio throughout, the pyramid's final height would have equalled the base centrepoint-corner dimension of 255RC. (360.6 ÷sqrt2 = 255)

For whatever reason part way through construction, the angle of slope was changed. In order to continue building and maintain an accurate shape, a new height:centrepoint-corner ratio had to be calculated to provide the means for the corner edges to meet at a new virtual apex.

Effectively to build a new pyramid, with a base formed on top of the completed lower section.

The height chosen to make the change was when the lower section had reached a significant height of 90RC. At this height the square top of the partially completed pyramid, has corners placed 165RC from the centrepoint (255-90=165).

This dimension, divisible by 3, allowed a new height:centrepoint-corner ratio of 2:3 to be calculated and applied using the virtual apex method from this height onwards.

The 'new' pyramid, a further 110RC tall (165÷3X2=110), fits exactly on top of the lower section, with each face sloping at an angle of 43º20'

A reference core structure would be required to complete this section, with each of the reference heights leaving a height **remaining** to the 'new' apex divisible by 3 in order to provide a straightforward means of calculating the pyramid outer corner positions.

These are the dimensions and slope angles of the Bent Pyramid and what is quite clear from its diagonal cross section dimensions is that they provide an elegant and straightforward construction process for a dual shaped pyramid of height 200RC.

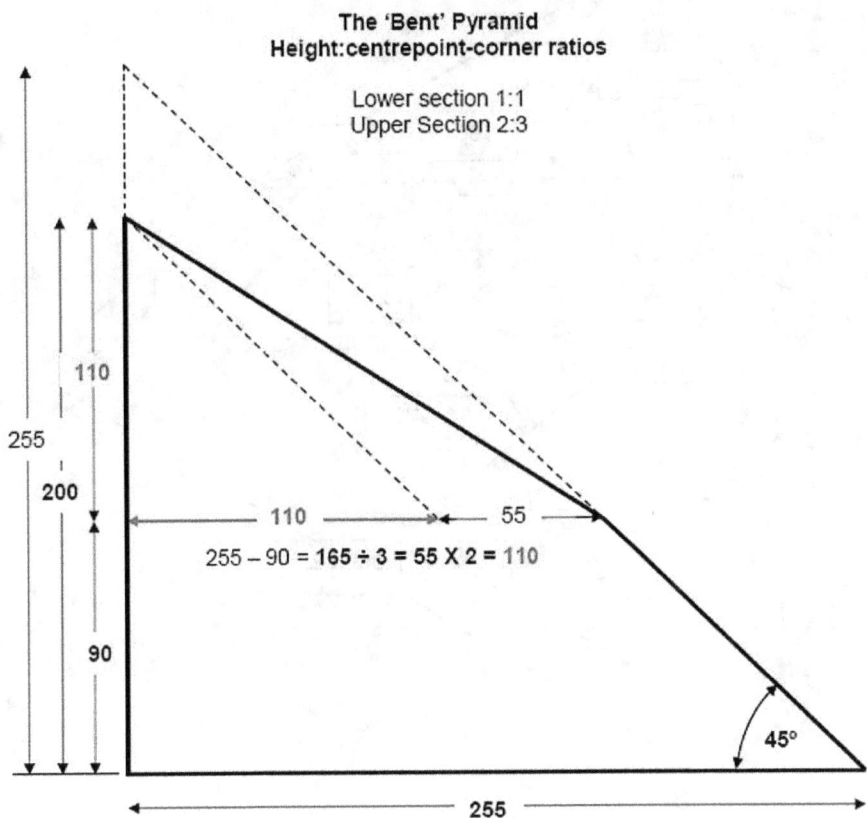

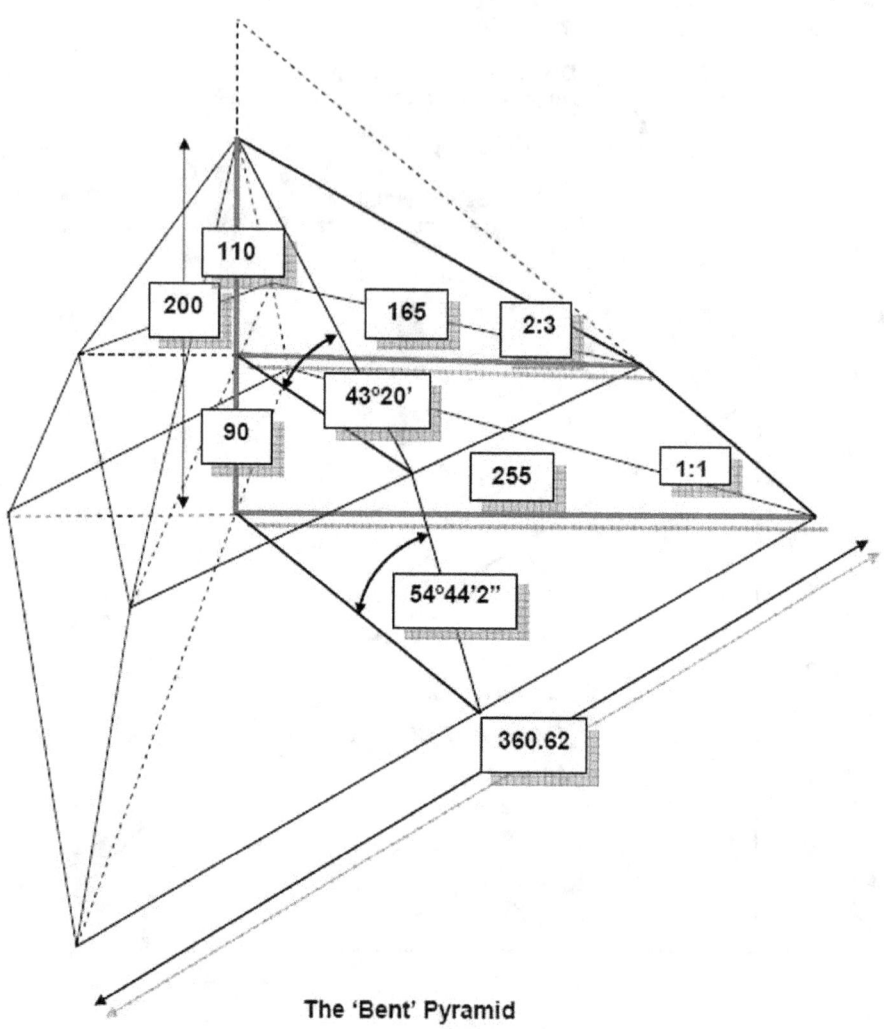

The 'Bent' Pyramid

Height:centrepoint-corner relationships for Upper and Lower sections creating face slope angles and side length as found

Anomalies?

It has been reported that the sides of the lower section of this pyramid are slightly concave where they connect to the upper section and their corners also indicate alignment problems at the same point.

A possible explanation for this might be that errors were made in positioning the base corners, or the corner edge was rising off line, which would then be perpetuated as construction continued, due to a demand that they ran straight. As the pyramid was built over an existing structure, it would have to at least equal its height before errors would be noticed and any remedial action taken.

If it was found that the centrepoint-corner dimension did not equal the intended base centrepoint-corner dimension minus the height built, or the top wasn't square, it would be recognised that the corner edges would have to be modified in their direction at some point if they were to meet at a single point, or at a specific height.

It might have been here that a decision was taken to continue with the corner edges running straight and then make a severe change at 90RC in order to resolve this problem. The concavity found in the lower faces might therefore signify their attempt to connect this section to the accurately measured 'base' of the upper section at 90RC, with its centrepoint-corner dimension of 165RC (255-90).

Petrie's survey has described how the boundary wall surrounding this pyramid has sides close to 560RC and therefore a virtual centrepoint 396RC from each corner (560X70/99). If the boundary wall position was in reality a reference square, then new corners placed a further 141RC inside its partial diagonals, creates centrepoint-corner dimensions of 255RC.

This base square has side lengths of 360.62RC and a distance to the boundary wall of approximately 100RC, which might confirm that this was the surveyor's real intention (141÷√2 = 99.7 and 560-199.4 =360.6). The dimension chosen might indicate the original intentions of the builders; 255RC suggests a pyramid with a single slope of 54º44'2" to a height of 255RC, rather than a dual sloped pyramid as found.

As the virtual apex method does not require the base sides of a pyramid to be measured, it would explain why this pyramid and others, including the Great Pyramid, do not have sides either of equal length, or dimensions which are whole numbers of Royal Cubits.

The 'Bent' pyramid was the first to be completed with flat faces and its dimensions show that the virtual apex and virtual centrepoint methods would have provided a simple and reliable means for building it to the final design, regardless of whether the dual shape was intended or was a response to errors during construction.

The diagonal dimensions found; of the boundary wall, the pyramid base diagonal, the diagonal at 90RC, and the overall height; each relate one to another in a uniquely straightforward and easily understood form. In addition to the simplicity of the diagonal dimensions, it can be noted that the two slopes in the profile of this pyramid do not provide whole number height to base proportions, vital for a 'rise and run' construction process, or a regular seked.

This would confirm that the pyramid profile was not a factor in the construction process itself and might suggest it had no significance in the pyramid's design.

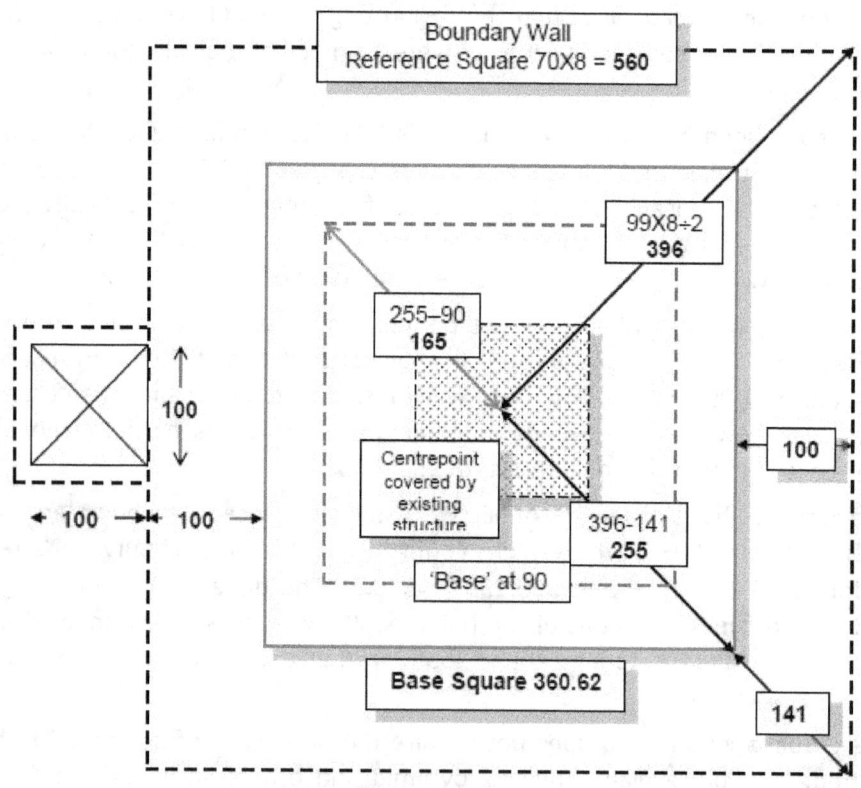

The 'Bent' Pyramid

Plan view showing the dimensions of the Boundary Wall and how its use as a Reference Square might have enabled the builders to form the pyramid base with a known dimension to the covered centrepoint

Dimensions in Royal Cubits

The Northern 'Red' Pyramid

This pyramid was constructed immediately after the Bent Pyramid and its shape repeats that of its upper section. Its face slopes have been determined as 43º20', based on a surveyed height of 200RC and base sides of 424RC. These dimensions and corresponding slope angles would result from employing a height:centrepoint-corner ratio of 2:3 throughout construction, to provide a virtual apex at 200RC to which the corner edges would aim.

The centrepoint to corner dimension at the base of 300RC would tend to confirm that the diagonal was a significant element in the design and once again it can be noted that this profile does not provide a whole number seked. Those advocating the 'rise and run' method of shaping a pyramid have also noticed and have responded by suggesting that this pyramid might instead have 'settled' and as a result its actual slope was close to 45º (or a seked of 1:1). They say that new data indicates that this pyramid might have had a height of 210 RC, which is 10RC more than originally calculated and a side length closer to 420RC. A rise and run of 1:1 could then have been employed to create the shape.

However, if this new survey is correct then the chosen dimensions and ratio, used to form this shape, might also provide confirmation of the Ancient Egyptian's knowledge of the relationship between the side and diagonal of certain squares. A pyramid with a profile of 1:1 also has a diagonal cross section height:centrepoint-corner ratio of 1:√2. (or 1:99/70 or 70:99)

A height:centrepoint-corner ratio of 70:99 can therefore be readily applied to form this shape, as the height is 210RC or 70 X 3 and the centrepoint-corner dimension 297RC, or 99 X 3.

The ratio can be applied, at those heights, which leave a remaining height to the apex in any combination of 70RC or 99RC, as each will allow the dimension from the centrepoint to the corner to be easily calculated.

The first reference height could therefore be 12RC, as at this height 198RC remains to the apex (99 X 2) and the corner positions are exactly 280RC (140 X 2) from the centrepoint.

Subsequent heights are 41, 70, 111, 148 and 198RC, with corner positions 239, 198, 140. 99 and 17RC from the centrepoint, respectively.

It is clear that whichever of these two shapes were intended, applying the virtual apex method, could have formed either, or both.

A Corbelled Ceiling within the Red Pyramid

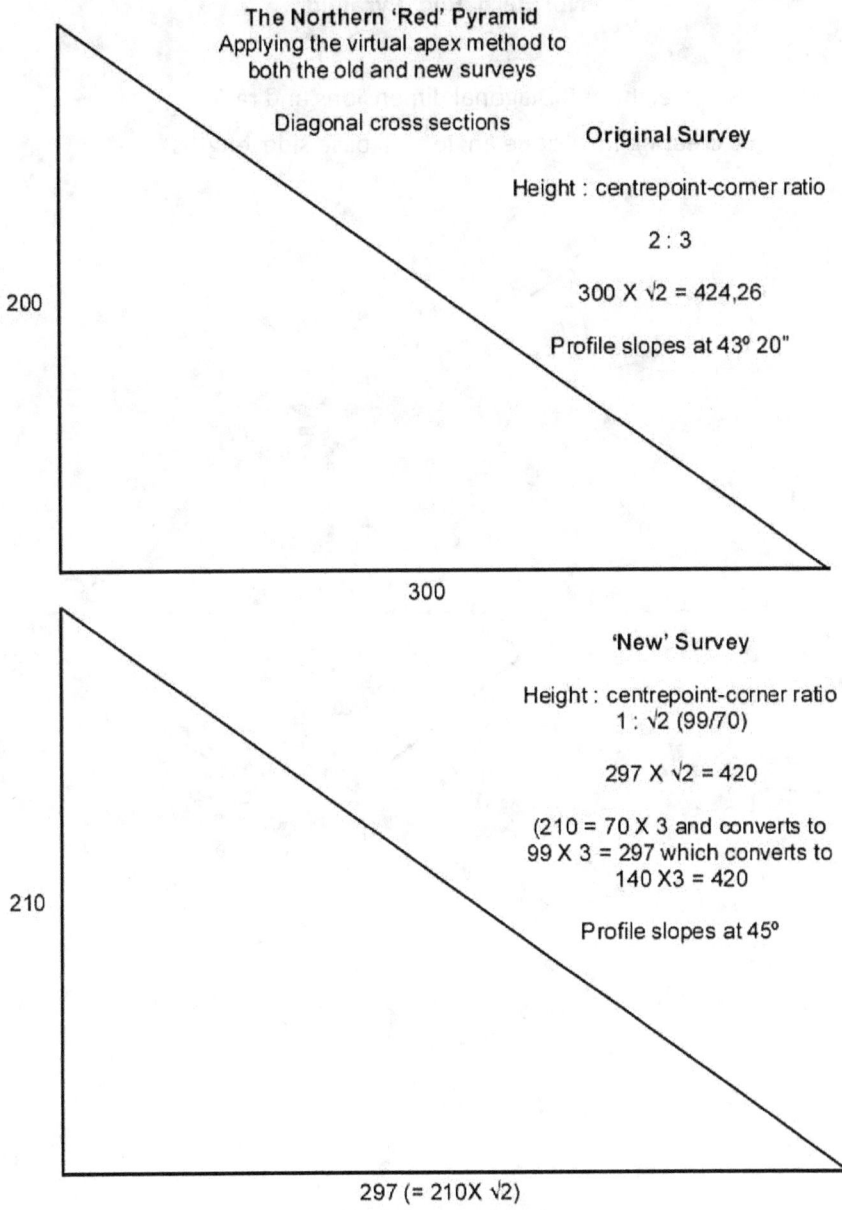

Northern 'Red' Pyramid

Height and Diagonal dimensions and ratio creating face slope angle and base side lengths

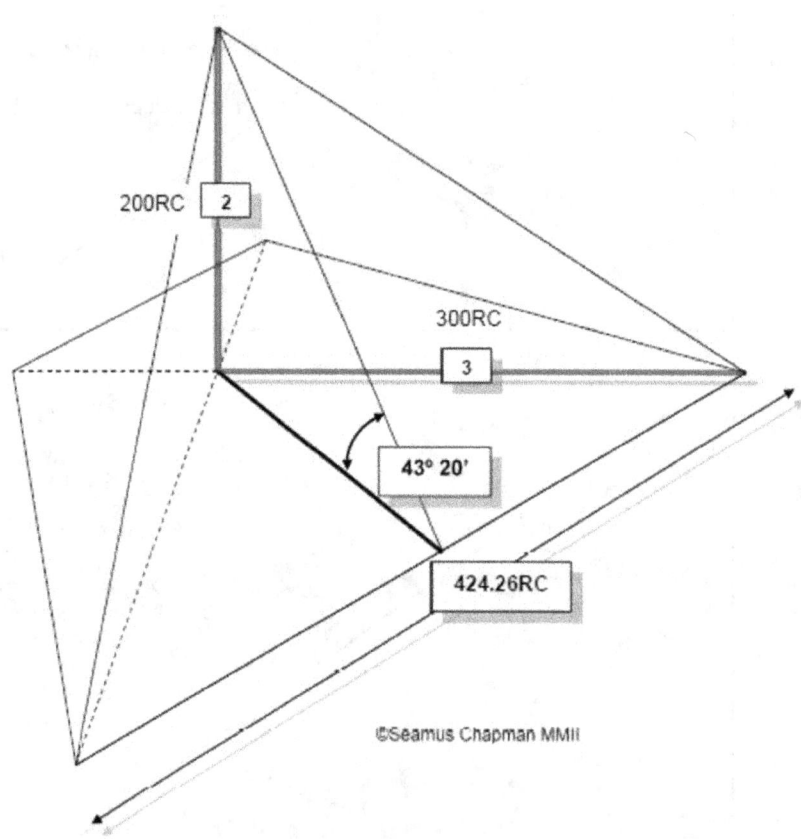

The Pyramid at Meidum

The pyramid at Meidum might have been used as a test bed for pyramid construction technology and a study of the special conditions, which applied could indicate the knowledge the Ancient Egyptians had of pyramid geometry.

Originally constructed as a stepped structure it was abandoned during the period when the Bent and Red Pyramids were constructed. When the builders returned some 15 years later, it was enlarged by adding additional steps and then a smooth outer casing.

If the intention was to make a smooth sided pyramid, why was it necessary to add the extra steps?

In order to make a pyramid accurately it is necessary to identify the centrepoint at various heights. Because the original stepped structure was in the way and the centrepoint covered, the builders were unable to confirm this directly and had to rely instead on a virtual centrepoint throughout. The extra stepped stonework completed to selective heights and dimensions, could have provided the virtual centrepoint-corner dimensions simply by measuring along their sides. In this case it would confirm that the Egyptians had to be aware of the relationship between squares with sides of 70 or 99 having diagonals of 99 or 140 and applied this knowledge when building their pyramids. If the Meidum pyramid had been completed its angle of slope, based on remaining casing blocks, would have been the same as the Great Pyramid and a study of its other dimensions suggest that the same height:centrepoint-corner ratio of 9:10 could have been used to construct it, in this case creating a base side length of 275RC and a height of 175.5RC. During construction, the side lengths of each of the extra steps would provide a means to calculate a known virtual centrepoint-corner dimension by using multiples or combinations of 99 and 70. The heights at which both this and the virtual apex formula could be applied easily would be each of those, which left a height to the apex divisible by 9.

Heights, which leave a height **remaining** to the apex of 144, 126, 108, 90, 72, 54, 36, and 18RC, with stepped core side lengths based on 70 or 99RC, would have readily provided a means to calculate the dimension from the virtual centrepoint, to the pyramid outer corner. At these heights the pyramid corners are 160, 140, 120, 100, 80, 60 40 and 20RC from the virtual centrepoint, respectively. For example, on a step at a height of 31.5RC, a square with sides of 198RC, has corners 140RC from the virtual centrepoint. The pyramid outer corners are a further 20RC along its extended diagonal.

$$175.5 - 31.5 = 144 \div 9 = 16 \times 10 = 160$$

It is unusual in Ancient Egypt to find two major structures with the same shape. Perhaps the return to this pyramid would be partially explained if it had been used to provide information about the design and construction processes, using a virtual apex and centrepoint to corner ratio of 9:10, which could then be applied in building the Great Pyramid, which followed.

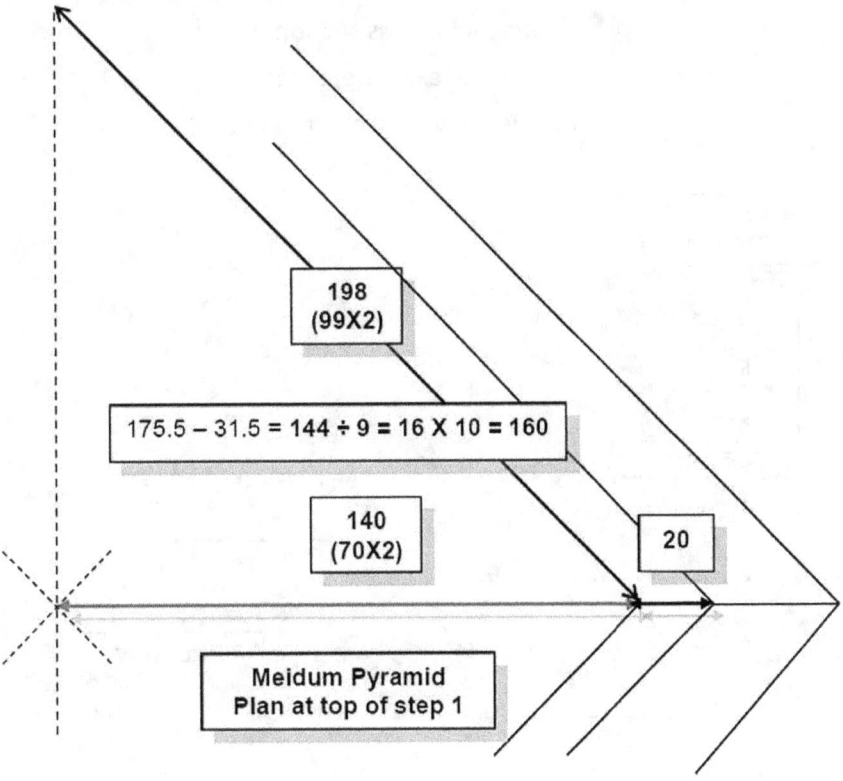

Side length of a step used to determine a centrepoint-corner dimension and the position of the façade corner

Meidum Pyramid
Diagonal Cross section
How the 'extra steps' can
provide the virtual centrepoint

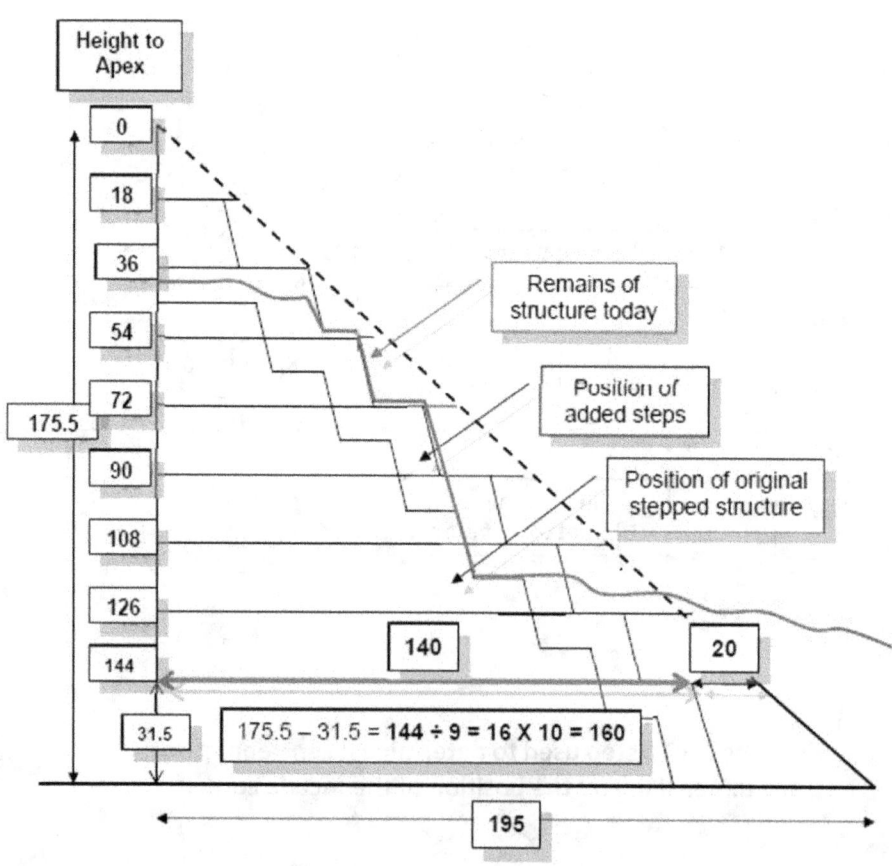

Meidum Pyramid

Height and Diagonal dimensions and ratio
creating face slope angle and base side length
(Ratio same as Great Pyramid)

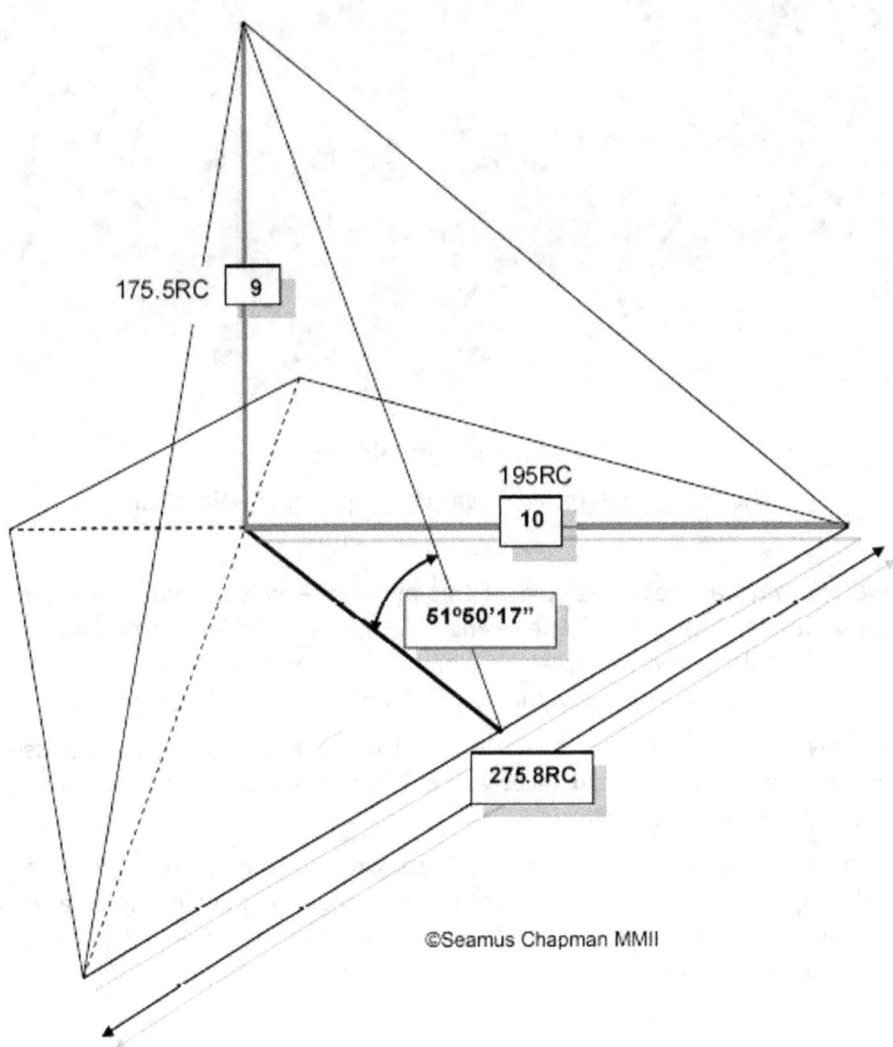

©Seamus Chapman MMII

The Meidum Pyramid Casing

The Meidum Pyramid – Was it Completed? – Did it Fall?

(or was it pushed?)

There is always a question asked of this pyramid – was it completed or did it collapse during construction? The casing shown here provides a logical answer. In order to install the facade, access is from a platform and ramp structure which surrounds the whole exterior of the pyramid and rises as each course is laid.

It is only when the façade has reached the apex that this is progressively dismantled and the pyramid faces are cut flat to the lines inscribed on their upper edge during installation.

The casing at the base of this pyramid is shown in **a finished condition** (despite weathering in parts), and as **this final effort** could only have occurred after the last elements of the ramp and platform system had been removed **the pyramid must have been completed.**

Whatever happened after that...

The Great Pyramid of Khufu

The cross section of this pyramid has a height of 280RC and base sides close to 440RC. This creates a profile described as 14:11 or seked of:

1RC : 5 palms, 2 fingers.

In order to apply the Virtual Apex method to build a pyramid with this shape, the diagonal length of the base square must be calculated and then compared with the height to provide a height to base ratio.

The procedure for calculating the base corner to centrepoint dimension in a square with sides of 440RC and using knowledge of a reference square and long division might have taken the following course:

440 ÷ 2 = 220 = 70 X 3 + 10 (70 palms)

Therefore the diagonal length from the base corner to the centrepoint is:

297 + 99 palms = 297+14 +1 palm = **311RC+1palm (311.11rRC)**

The height:centre-point-corner ratio at the base is therefore:
280RC : 311RC+1palm

The next step would be to determine a height:centrepoint-corner ratio, which when applied will allow the corner positions at various heights to be found routinely and ensure that when connected the pyramid corner edges meet at a single point at the chosen height of 280RC.

The closest ratio is 9:10 as 279/9 X 10 = 310

If this height:centrepoint-corner ratio is applied according to the virtual apex method, it would create a pyramid with a profile close to 28:22. In this case the pyramid would have a height of 279 and side length 438.4.

If the pyramid must have a height exactly divisible by 28, then the most reasonable action would be to add 1RC to both the centrepoint-corner dimension and the height and this is what the Ancient Egyptians appeared to have done. The 9:10 ratio can be still be routinely implemented during construction, to achieve these dimensions, by choosing reference heights where the **remaining** height to the apex is exactly divisible by 9 and multiplying the product by 10, to give the centrepoint to corner dimension at each height.

As the chosen height was to be 280RC then these heights would be 10, 19, 28, 37 etc.

For example at a height of 28RC, the height remaining to the apex is 252RC; therefore the pyramid corner at this height is 280RC from the centrepoint.

(280 – 28 = 252 ÷ 9 = 28 X 10 = 280)

Evidence for these actions can be found in the dimensions and shape of this carefully measured pyramid. The average side length is 439.82RC, which might confirm an intention to place the corner positions 311RC from the virtual centrepoint, but not necessarily a measured 440RC apart.

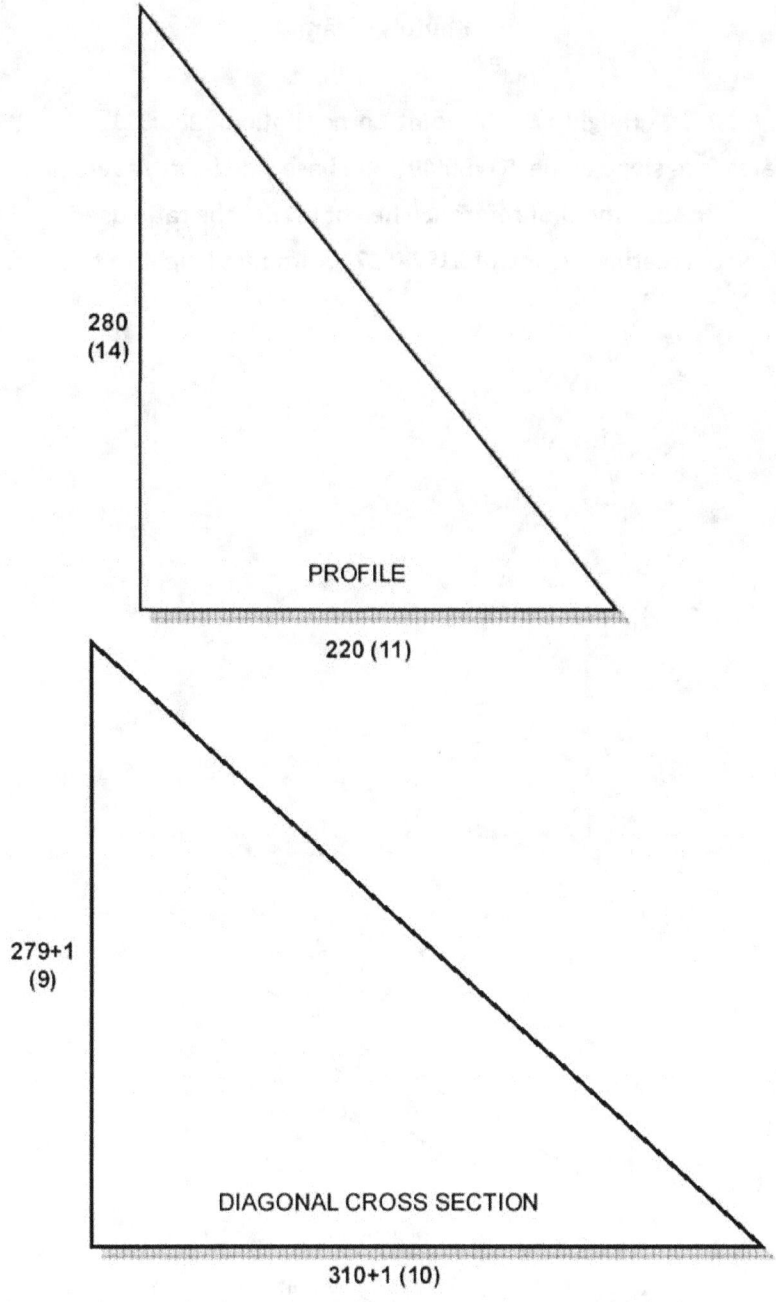

Significant Dimensions of the Great Pyramid

Khufu's Pyramid

Height : Centrepoint-Corner Ratio of 280:311
creates face slope angle 51º50'40" and base side length averaging 439.82
Above the first reference height (37RC) the ratio used is
9:10 creating a slope of 51º 50'17" from that height to the apex

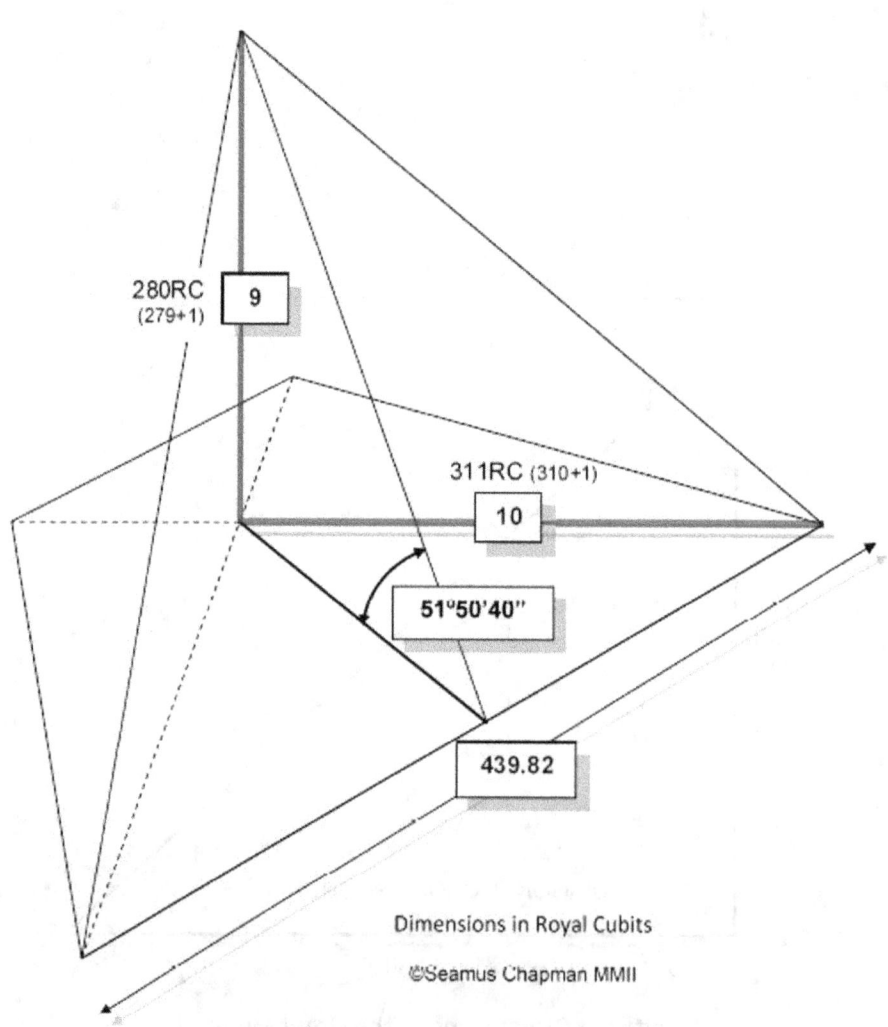

Dimensions in Royal Cubits

©Seamus Chapman MMII

Great Pyramid Boundary Wall Foundation Line

Great Pyramid East Side

Placing the base corners 311RC from a virtual centrepoint can be readily achieved by measuring inwards from the partial diagonals of a larger reference square, with a known dimension from the virtual centrepoint.

For example, a square with sides of 495RC (99 X 5) has a known diagonal of 700RC (140 X 5) and a centrepoint to corner of 350RC. Measuring 39RC along the partial diagonals of this square provides corner positions 311RC from the virtual centrepoint.

The foundation of the boundary wall of Khufu's pyramid forms a square with sides of 495RC.

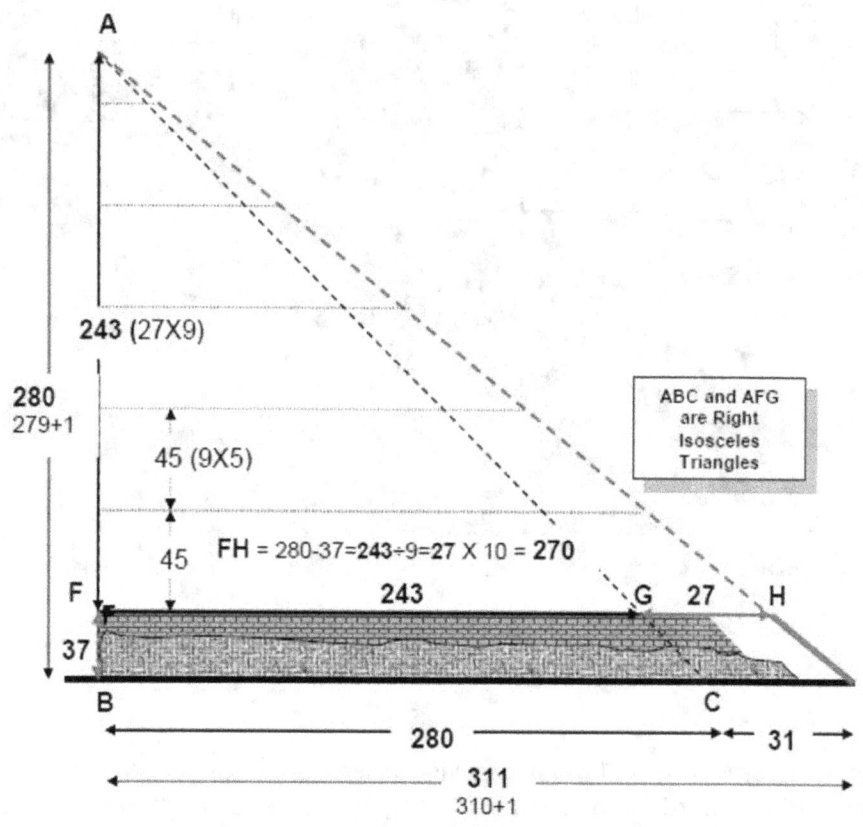

Diagonal Cross Section Khufu's Pyramid

© Seamus Chapman MMII

The core structure is taken to a reference height of 37RC and covers the 7 metre high central mound.

The prefabricated Queen's Chamber is assembled on the centreline as the façade is added to the same height

Khafre's Pyramid

Top of Khafre's Pyramid with casing still in place
Note the accuracy of the corner edges

The cross section of this pyramid has dimensions close to height 274RC and base 411RC, which suggests a profile of 28:21, or 3 – 4 – 5, or a seked of 1RC : 5 palms, 1 finger.

In order to determine a height : centre-point-corner ratio in a pyramid with this profile, additional steps are necessary in applying the reference square formula, in the following way:

A pyramid 280 high, with a height to half base profile of 4 to 3, would have a base side of 420 and the centre-point-corner dimension in this case 297, as:

$$420 \div 2 = 210 = 70 \times 3, \text{ which converts to } 99 \times 3 = \mathbf{297}$$

A pyramid 264 high, with a height to half base profile of 4 to 3 would have a base side of 396 and the centre-point-corner dimension in this case, 280, as:

$$396 \div 2 = 198 = 99 \times 2, \text{ which converts to } 140 \times 2 = \mathbf{280}$$

The midpoint of these two examples is a pyramid with a height of 272 and a base side length of 408 having a virtual centrepoint:corner dimension at the base of 288.5 as:

$$280 + 264 = 544 \div 2 = 272 \text{ and } 420 + 396 \div 2 = 408.$$

$$\text{And } 297 + 280 = 577 \div 2 = 288.5$$

There are two whole number height:centrepoint-corner ratios based on a height of 272 and a centrepoint-corner dimension of 288.5, which are necessary for applying the virtual apex method in its simplest form.

These are:

$$17:18 \text{ as } 272 \div 17 = 16 \times 18 = 288.$$

And $\quad 16:17$ as $272 \div 16 = 17 \times 17 = 289$

These are the only height:centrepoint-corner ratios which will provide a means to actually construct a pyramid with a profile close to 28:21, a seked of 1RC to 5 palms 1 finger or 3 - 4 – 5.

As two ratio options were available, we can compare the carefully measured profile of the pyramid with each, in order to confirm which might have been applied.

Khafre's pyramid has a measured slope of 53°10'.

A pyramid with an exact 28 : 21 or 3 – 4 – 5 profile has a slope of 53°7'.

A pyramid with a diagonal cross section of 16:17 has a profile slope angle of 53º5'.

A pyramid with a diagonal cross section of 17:18 has a profile slope angle of 53º10'.

This might confirm that of the two ratio options available 17:18 was chosen.

In this case, the corner positions can always be accurately found at those heights, which **leave a height remaining to the apex** exactly divisible by 17 and then multiplying the product by 18, to provide the dimension from the centre-point to the corner. As the height of Khafre's pyramid is 274RC, then the ideal heights above the base to place the corner edges, would be 19, 36, 53RC etc. For example at a height of 53RC, 221RC remains to the apex. The corner edge at this height is therefore 234RC from the centre-point as:

$$274 - 53 = 221 \div 17 = 13 \times 18 = 234$$

As the overall height was to be 2RC taller than the calculated height for accurately controlling the shape, the centrepoint-corner dimension of the base square was also increased by 2RC, from 288.5 to 290.5RC giving a side length of 410.8RC

The corner positions could have been found by measuring inwards from a larger reference square, with a known centrepoint-corner dimension, as described for Khufu's pyramid, although it will be shown later that the SE corner could have been placed by using a dimension taken from a reference diagonal covering the whole Giza site.

The additional height does not affect the construction process and in fact evidence for exactly this procedure in this pyramid, can be found in the dimensions and slope angle found.

Khafre's pyramid has a base course of granite blocks 2RC tall. If the dimensions of the pyramid are taken from the top of this course, its height is 272RC and its centrepoint-corner dimension 288.5RC and therefore a side length of 408RC.

In this case as 272 is divisible by 4 and 204 by 3, a pyramid profile based on this height and base would more clearly demonstrate a 3-4-5 relationship, if this had been an intention of the Ancient Egyptians.

It might also be the case the design of both the Khufu and Khafre pyramids was based only on the diagonal cross-section dimensions, vital for implementing the virtual apex method, with the profile dimensions and relationships only an automatic consequence.

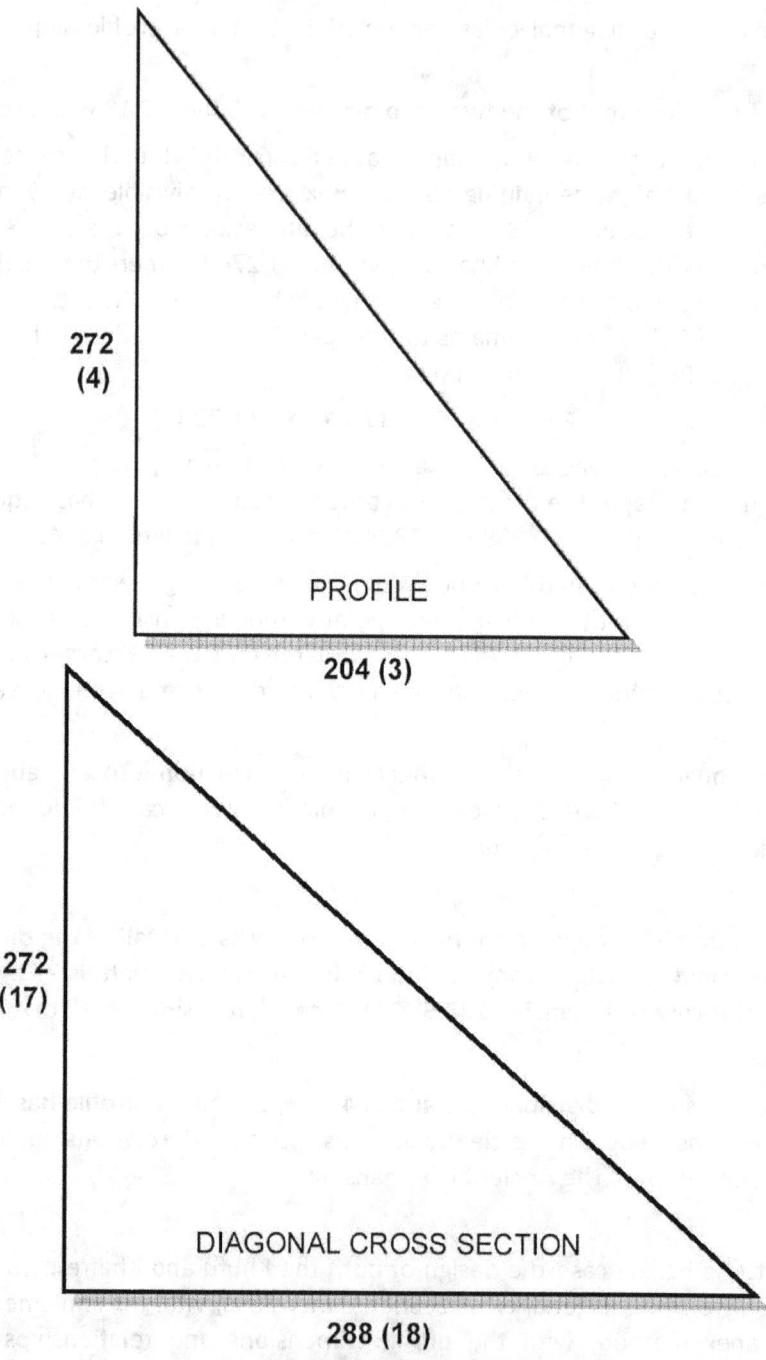

Significant dimensions Khafre's Pyramid
(From top of granite course 2RC tall)

Khafre's Pyramid
Height : Centrepoint-Corner Ratio creating face slope angle and base side length
(From top of granite course 2RC tall)

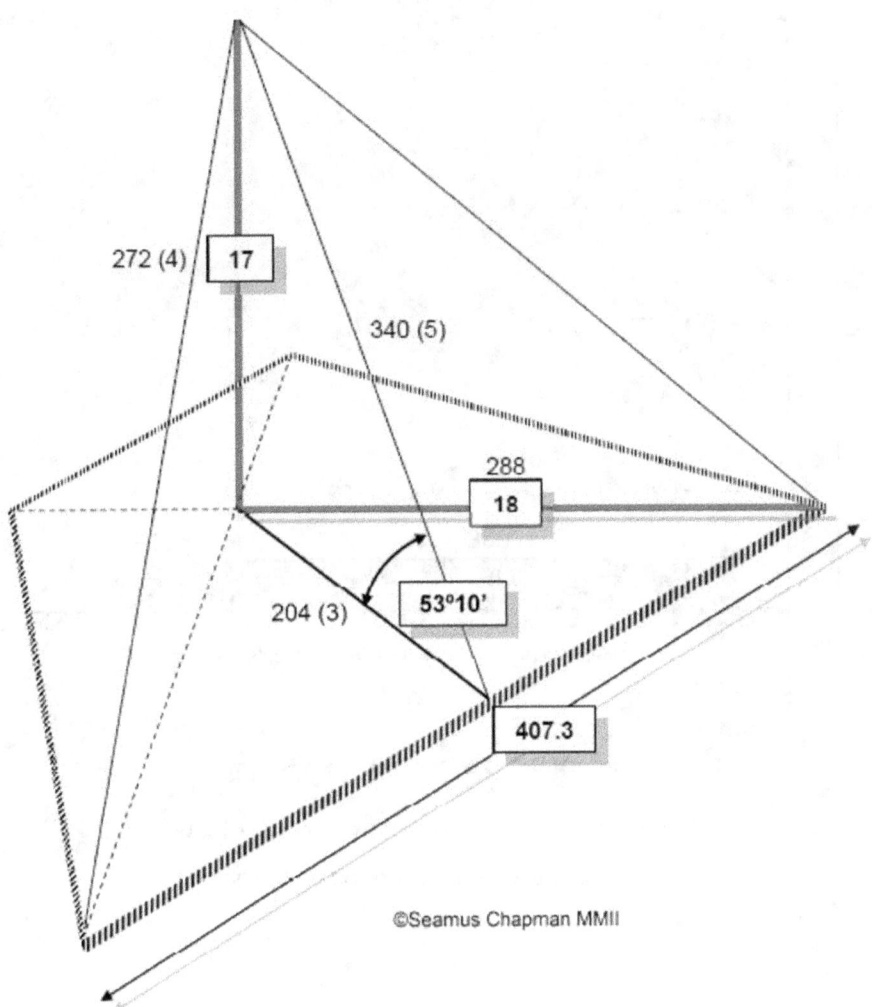

Dimension in Royal Cubits

©Seamus Chapman MMII

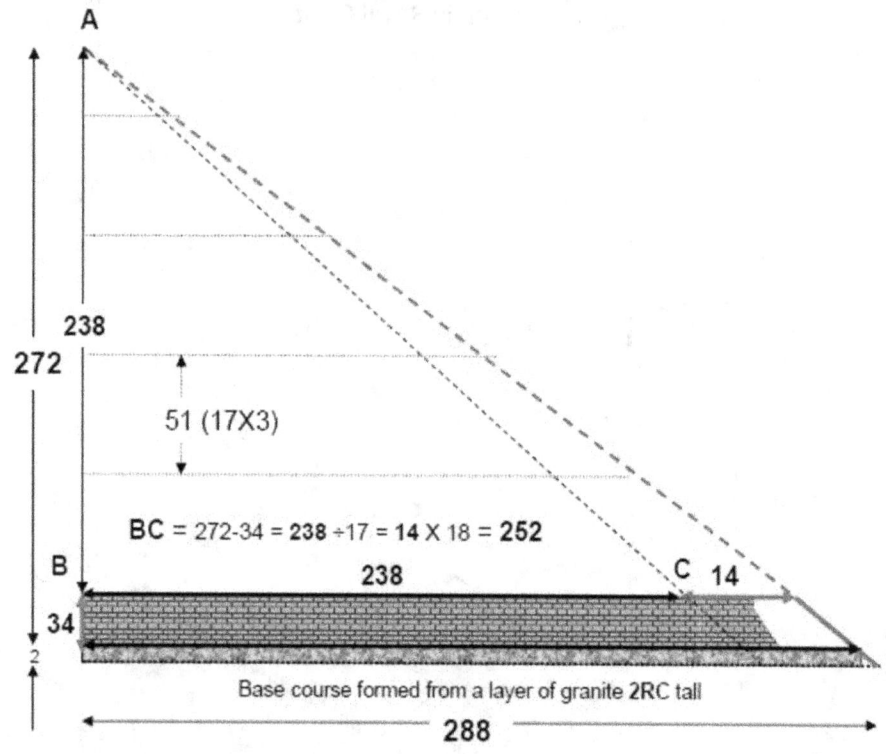

Diagonal Cross Section Khafre's Pyramid

**Showing the core at a reference height of 34RC
above the single granite base course**

Menkaure's Pyramid

This pyramid is the smallest of the three at Giza, being only one tenth the volume of the Great Pyramid. As its base is not clearly defined a number of assumptions were made by Petrie, which included the intended height of its pavement and the point at which it would have connected to the unfinished granite base blocks. The overall height was determined from this upper base dimension, using a slope angle of 51º10', which itself was a mean of angles found on facing blocks both on and off the pyramid (slopes ranging from under 51º to almost 52º) and surveyed slopes from different positions. He concludes the pyramid has base sides of different lengths, with an average close to 201RC and a height close 125RC, if the pavement had been installed.

Had the survey been carried out with knowledge of the virtual apex method which does not use the face slope for accurate construction, but a simple height : centrepoint-corner ratio, then this could also have been considered as part of the surveying parameters and it would be reasonable to compare the range of possible dimensions with the demands of its geometry.

In this case, the intended dimensions of this pyramid might have been presented as a base centrepoint to corner of 144RC, creating an unmeasured base side length of 203.6RC and height 126RC. They are within the margins of error of the original survey and more accurately describe the dimensions of and from the bedrock base. It also gives a base diameter which is half that of Khafre's pyramid taken from above its granite base course.

The height of 126RC is divisible by 7 and application of the virtual apex method using a height : centrepoint-corner ratio of 7:8 (centrepoint-corner dimension at base = height 126 + 1/7 = 144), will provide the means to form a pyramid accurately to these dimensions and in this case if applied with 100% accuracy its faces would slope at 51º4'. The first action of the pyramid surveyors would be to form a reference square with sides of 210RC and corners known to be 148.5RC from its centrepoint. The corners of the base square are then marked 144RC from the virtual centrepoint.

It is the perimeter of this square that the base course of unfinished granite casing blocks would have been and are found to be, aligned. Any errors found in the dimensions of this square might therefore have resulted from inaccurate positioning of the reference square corners and not from inexact measuring of the base sides.

It can also be noted that a clear seked is not provided in the profile of this pyramid, even though 126 is also divisible by 28. However a seked of 1RC to 1RC+1palm (28:32) is formed exactly by the corner edge and whereas the diagonal cross section can be presented as height 7 to half-base 8, the profile presents a triangle of height 7, base 8÷√2 and hypotenuse 9.

Menkaure's Pyramid – Pavement level

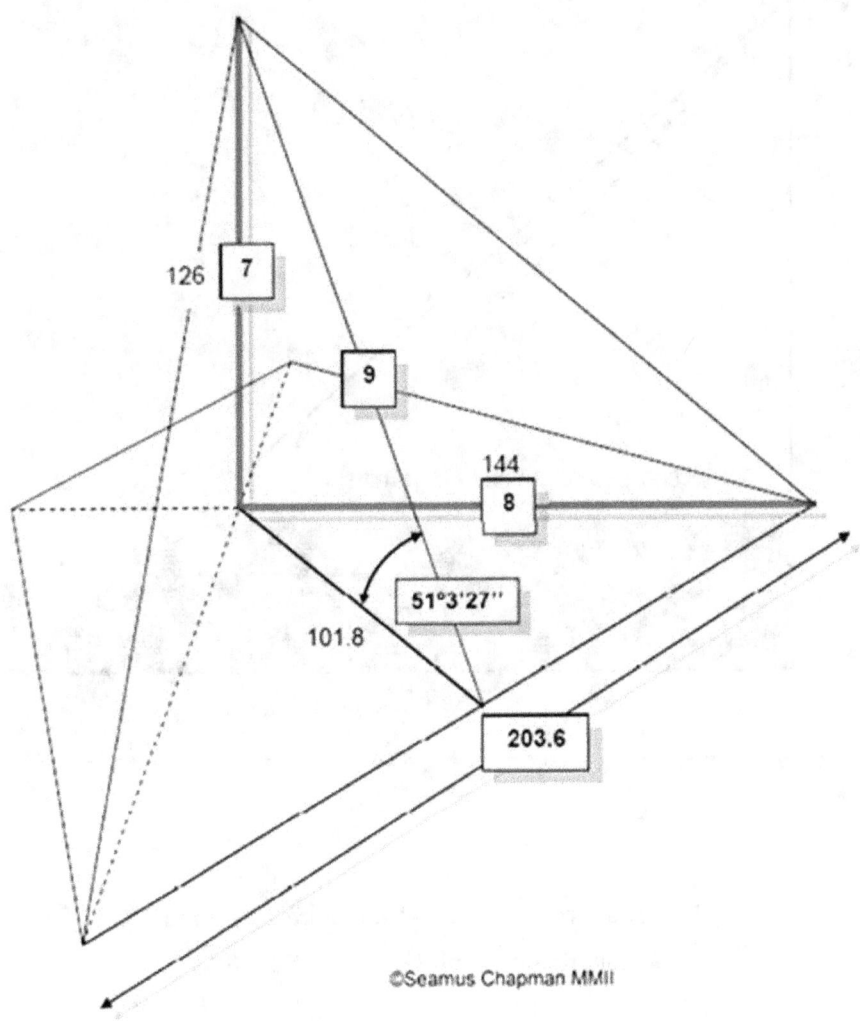

Dimensions and proportions of Menkaure's Pyramid

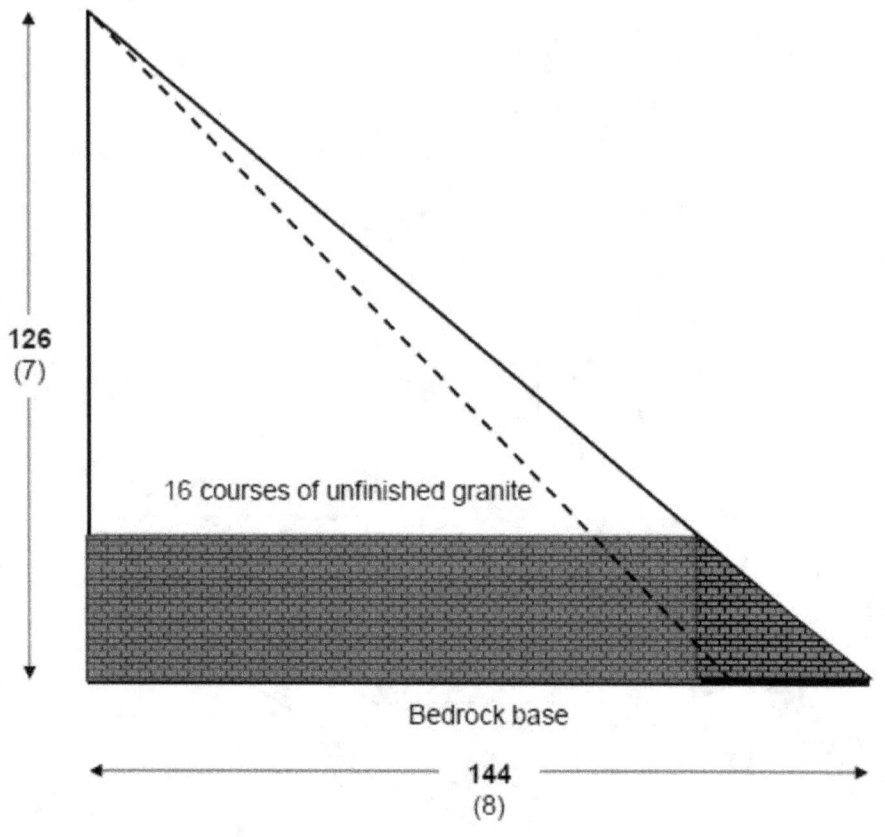

Diagonal Cross-Section - Menkaure's Pyramid
Height:centrepoint-corner ratio 7:8

Corner Edge Seked 1RC:1RC + 1 palm

©Seamus Chapman MMII

Pyramid height:centrepoint-corner ratios and their profile slope angles

Ratio	Angle	Notes
1:1	54°44' 10"	Bent pyramid - lower section (measured average)
2:3	43°18' 50"	Bent pyramid - upper section
2:3	43°18' 50"	Red pyramid
3:4	46.686°	
4:5	48.527°	
5:6	49.685°	
6:7	50.479°	
7:8	51°3' 25"	Menkaurie (my observation from bedrock base)
8:9	51°29'52"	
9:10	51°50' 17"	Meidum (forming 'seked' 28:22)
9:10		Khufu (280:311 or 279+1 and 310+1 gives a profile of 51°50'40")
10:11	52.124° 52° 7' 26"	Djedefre?
11:12	52.355°	
12:13	52.548°	
13:14	52.712°	
14:15	52.852°	
15:16	52.974°	
16:17	53.082°	
17:18	53°10' 40"	Khafre (measured average)
18:19	53.263°	

Notes

1. The profile angles shown above are derived from the diagonal cross-section proportions of symmetrical pyramids which can be applied simply to construct a solid pyramid to these shapes using the virtual apex method. As every major Egyptian pyramid profile is included, it follows that any competing construction method should at least be able to demonstrate the same.

2. Khufu's pyramid employed a 9:10 ratio, with 1RC added to the base centrepoint-corner and height dimensions to enable a final height of 280RC. (279+1:310+1). Khafre's pyramid employed a 17:18 ratio as it

provides a profile close to 4:3. This is supported if its dimensions are taken from the top of its single granite course, giving a height of 272RC; divisible by both 17 and 4.

3. The profile angles of some pyramids are often based on their height and base dimensions rather than direct measurement and some have attempted to make them fit to a whole number 'seked'; this being a requirement for a 'rise and run' construction process. In the case of the Bent and Red Pyramid's; as their dimensions do not provide a regular seked, some have suggested that they might have 'settled' or become 'distorted'. I have demonstrated elsewhere that it is impossible to build any pyramid using 'rise and run' as a construction process and maintain that the seked itself might have been invented much later as a simple device for making an estimate of a pyramid's height when only the base dimension and slope ratio was available for measurement.

4. As no pyramid has sides of equal length or a whole number of Royal Cubits, none can have a consistent angle in their profiles. Casing blocks have also been found with slopes which would fit other pyramids, demonstrating that the casing angle itself could not have been used to form the pyramid shape and must therefore be a consequence of a different alignment process.

5. **Diagrams showing the plan view of major pyramids, with elevation superimposed, to demonstrate how their profile slope or seked was automatically formed during construction.**

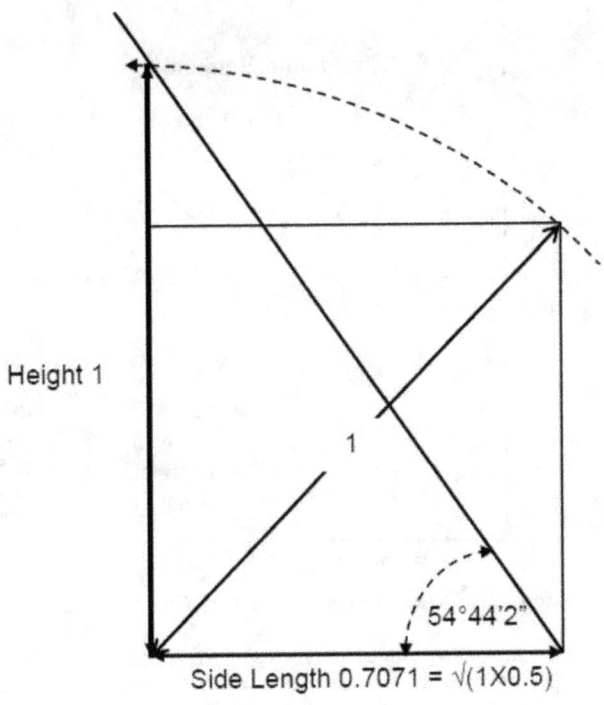

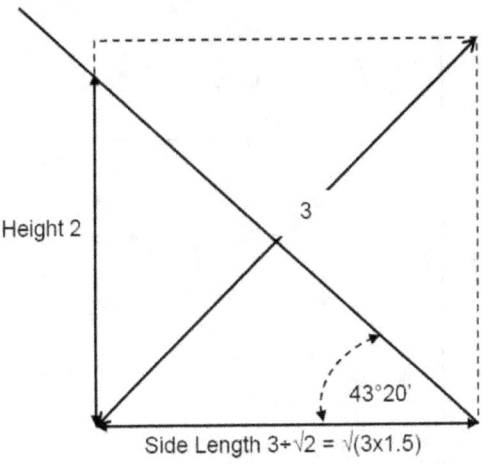

Face slope angles of lower and upper sections of the 'Bent' Pyramid

Derived from the diagonal cross-section Height : centrepoint–corner ratios
1:1 for the lower section and 2:3 for the upper section

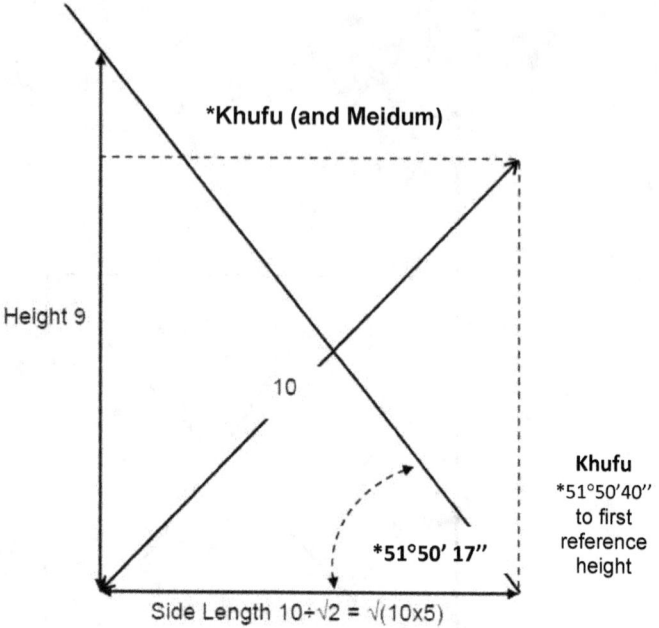

Face slope angles of Khafre's and Khufu's Pyramids

Derived from their Diagonal cross-section Height : centrepoint-corner ratio
Khafre's Pyramid 17:18 – Khufu's Pyramid 9:10 (and Meidum pyramid)

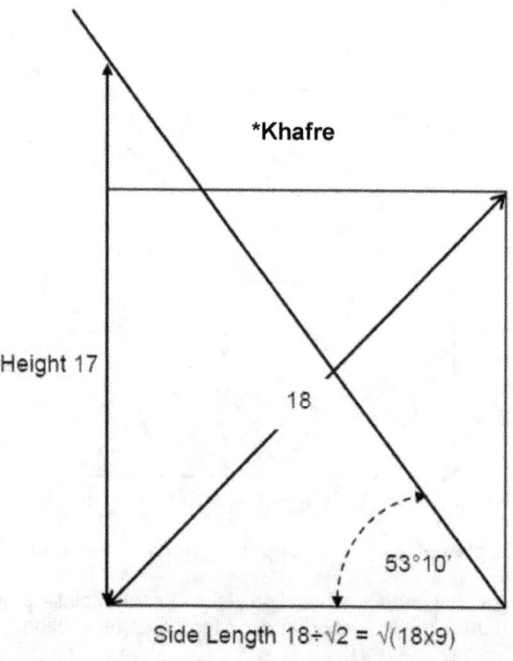

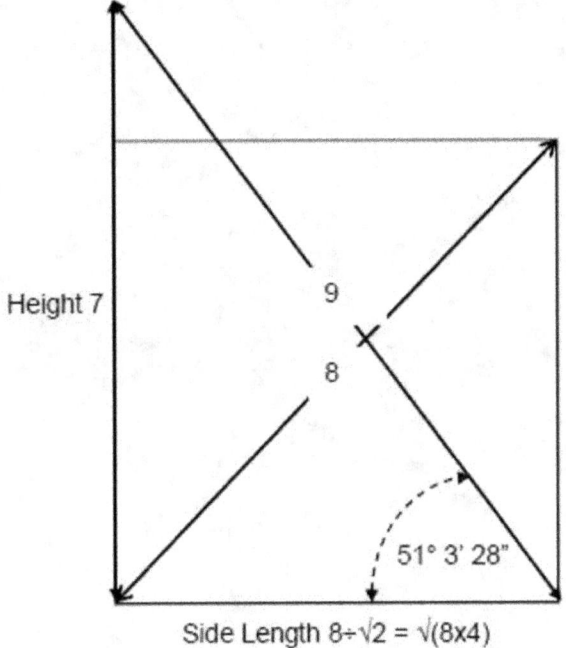

Face slope angle of Menkaure's Pyramid

Derived from its Diagonal cross-section
Height : centrepoint-corner ratio 7 : 8

Note: hypotenuse length 9

The Giza Site

Many attempts have been made to determine if there is a relationship between the dimensions and positions of the three pyramids at Giza. It is clear from reading the present work that the least we would expect from the Ancient Egyptians would be for them to have taken a practical and systematic approach to this task.

Legon has presented an interpretation of the origins of the dimensions found by Petrie and his diagrams clearly show the pyramid base dimensions and the spaces between them conforming to an overall site layout based on the diagonal or side lengths of large squares and rectangles with dimensions exclusively of 250RC, 500RC, 1000RC or 2000RC. However the topography of the Giza plateau would have prevented any of this geometry being used as the means for laying out the site, or for setting out the pyramid bases.

It might have been the case instead that this geometry was drawn out on a smaller scale elsewhere - say 1:100, and some of the dimensions taken from that.

As a consequence a more practical arrangement could then have been employed on site using a primary reference diagonal which would provide both a straightforward method for applying the geometry and all the dimensions and relationships found.

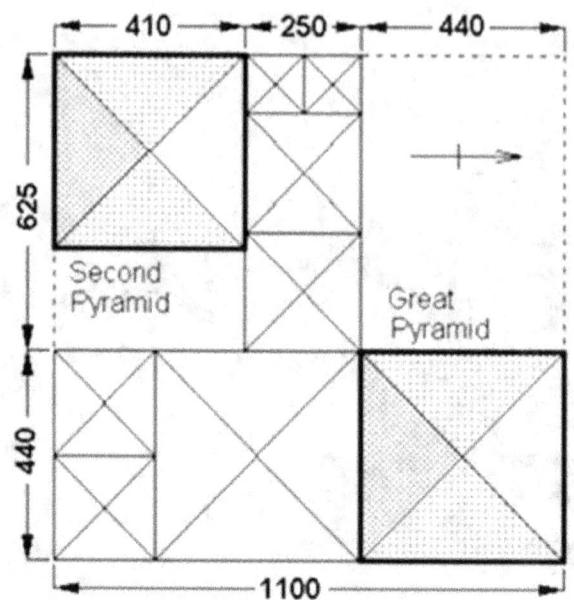

Modular Scheme Connecting the Second and Great Pyramids
Dimensions in Royal Egyptian Cubits

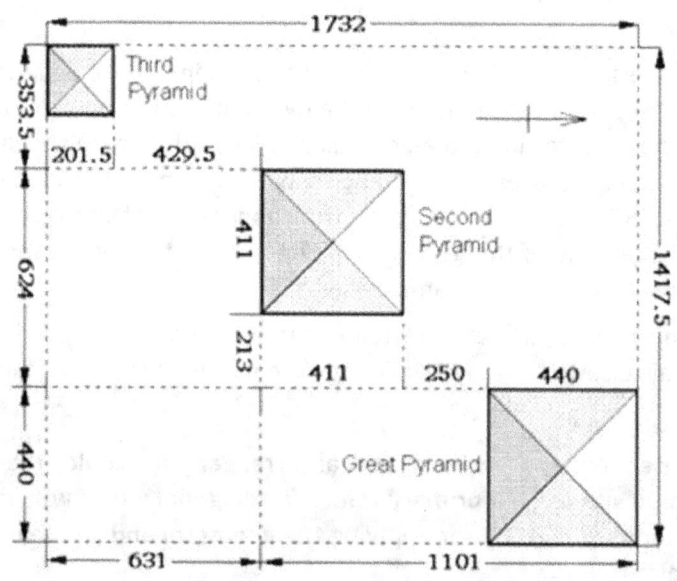

Dimensions of the Giza Site Plan in Royal Egyptian Cubits © J.A.R. Legon, 2000

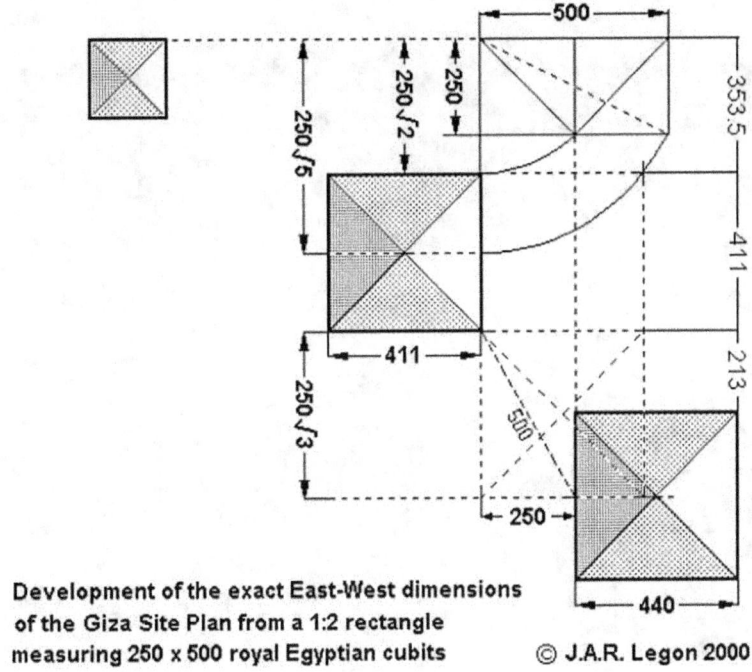

Development of the exact East-West dimensions
of the Giza Site Plan from a 1:2 rectangle
measuring 250 x 500 royal Egyptian cubits   © J.A.R. Legon 2000

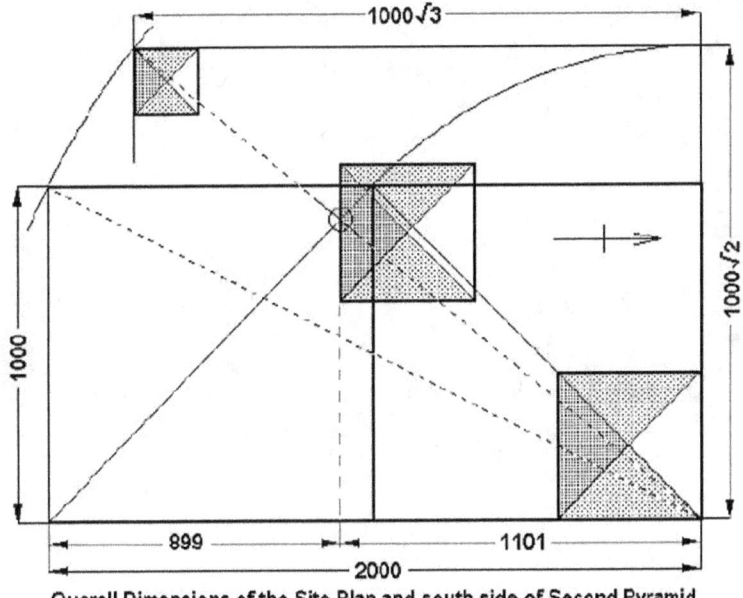

Overall Dimensions of the Site Plan and south side of Second Pyramid
Constructed from a 1000 x 2000 cubit rectangle   © J.A.R. Legon 2000

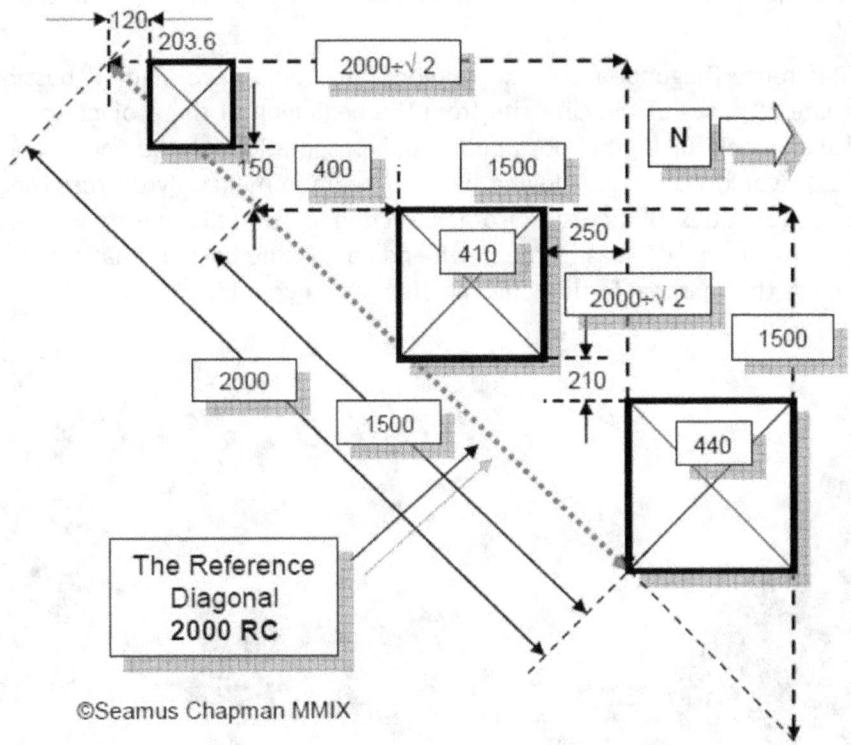

Layout and key dimensions of the Giza site based on
a Reference Diagonal of length 2000 Royal Cubits

The Reference Diagonal

A line 2000 Royal Cubits in length connected at 45º to the south-east base corner of the Great Pyramid, crosses what was the most level section of the Giza plateau and creates at its SW end a reference for the western side of Menkaure's pyramid and at its NE end a reference for the southern side of the Great Pyramid as both align to the sides of a square with this diagonal.

A line extended 400RC northwards from a point 1500RC from the NE end (or 500RC from the SW end), defines the SW corner of the base of Khafre's pyramid and the line of its western side. As this line is extended northwards it intersects with a line westwards from Khufu's southern side to form the sides of a square with a diagonal of 1500 RC and when extended to 1500RC intersects with a line extended westwards from the northern side of Khufu's pyramid. The references for Menkaure's base are more speculative as its base dimensions are not well defined. However if they are taken from the bedrock rather than the pavement level, its eastern side is 150RC west of Khafre's western side and the southern side is 120RC north of the SE end of the Reference Diagonal. This base square also has a diagonal of 288RC, exactly half that of Khafre's pyramid at the top of its granite base course.

The Reference Diagonal and its key dimensions would have made it possible to designate all areas of the Giza site from the beginning of the project, this being particularly helpful in the positioning and preparation of the many ancillary works. It would also have allowed work to begin immediately on removing the massive quantities of bedrock from a section of the site which would later become two levelled areas to the north and east of the base of Khafre's pyramid and to use this material for the building of Khufu's pyramid.

The western side of Khafre's pyramid showing the extent of site levelling around its base. A similar area exists along the northern side. Menkaure's pyramid is in the background

If this 200,000+ cubic metres of material, had not been routinely removed and consumed during Khufu's reign, it would have presented huge logistical problems later, including double handling when the base for Khafre's pyramid was being prepared close by. If this work only began on completion of Khufu's pyramid, where would the recovered material be stored until the preparation of the base platform for Khafre's pyramid was completed?

It appears the perimeter of the base of Khafre's Pyramid was determined at the top of the 2RC high granite course, which itself had to rest on a sub-base of other blocks at the southern side, due to the sloping nature of the site. In this pyramid also, the burial chamber is a simple rectangular space cut into the bedrock at ground level with a vaulted roof over. This space and its associated passage lined in granite could have been easily installed within the time available as the site excavation and base was being prepared.

Part I - Conclusions

The virtual apex method is a consistent and practical construction process for large pyramids, including all those in Egypt. It provides a physical means for defining a pyramid corner edge, which when connected will form a straight line to the apex at any chosen height. The accurate corner edge provides a straight line reference to which blocks forming the faces can be placed, in common with a universal building practice. It defines the shape of Khufu's, Khafre's and other pyramids exactly whereas other methods do not.

Repeated and careful measuring has failed to find any pyramid with a base side dimension which is a whole number of Royal Cubits or base sides of equal length, even though all casings still in situ show corner edges running straight to the apex.

This would result if chosen profiles had been modified to conform to the demands of providing a virtual centrepoint and virtual apex using diagonals, with base side lengths a consequence of this action and not a result of inaccurate measurement. In this sense all Egyptian pyramids can be considered as being constructed with the same degree of accuracy, if each employed the same method successfully in forming their shapes.

The whole number 70 – 99 side-diagonal relationship of a single reference square provides a simple means for calculating the diagonals of other squares and without knowledge of sqrt2. The corners of any square can be placed inside or outside the partial diagonals of the Reference Square, each with a known dimension from a virtual centrepoint. The Ancient Egyptians had been forming large squares for generations, which provided opportunities for empirically discovering this relationship and might explain why they chose to divide the Royal Cubit into 7 fractions. The Meidum pyramid might confirm knowledge of the reference square also, by explaining the purpose of the extra steps and how they might have been used to control the shape of the pyramid, when the centrepoint was covered throughout construction.

The virtual apex geometry and its application in Ancient Egypt for building pyramids can be summarised in the following way:

The formula for calculating the centrepoint-corner dimension of any pyramid and at any height, which if connected will meet exactly at the virtual apex is

<u>**Apex height – built height**</u>
Height:centrepoint-corner ratio

When using a virtual apex and virtual centrepoint to build an Egyptian pyramid, the chosen dimensions must provide a satisfactory whole number height:centrepoint-corner ratio in its diagonal cross section, so that the corner positions at specific heights can be easily calculated using basic arithmetic.

Whatever profile might have been chosen for an Egyptian pyramid, would be modified if necessary, to meet the demands of the diagonal cross section height:base ratio, described below.

The **height:centrepoint-corner ratio** used was always the height, plus a simple fraction of the height.

The **built height** would always leave a **height remaining** to the apex, exactly divisible by this fraction and be a **reference height.**

The centrepoint-corner dimension at each reference height therefore equals the remaining height to the apex, plus this fraction of the height and each of these two dimensions and their sum, would always be a whole number.

It can be noted that the same geometry can also describe the profile of a pyramid. However, at each reference height this would only provide a centreline-side dimension which has no function as a construction guide as the corners would have to be positioned last, contrary to a universal building practice.

The method can only be applied to the diagonal cross section to actually make a pyramid as this provides the four corner positions first, which when connected automatically forms a reference for the sides.

A core structure must be constructed to each reference height in advance of a facade, to physically provide a platform from which the upper corner positions can be placed and connected to the known corner positions below, thus defining each pyramid corner edge to that height. All partially exposed pyramids show evidence of a stepped core, with the exception of some later pyramids, which have a centrepoint-corner dimension equal to the height (Ratio 1:1), a ratio allowing corner references to be made at any height, without the need for a reference core.

The layout of the three pyramids at Giza appears to have been derived from the dimensions of large squares and rectangles with relevant setting-out points taken from a single reference diagonal 2000 RC in length. It provides the position of two sides for each pyramid's base and would also be vital for designating works in ancillary areas across the whole site from the outset, enabling the planned removal of material for use in constructing the reference core.

The Reference Diagonal must be in place before construction began, as it would have been impossible to connect a line to the SE corner of Khufu's pyramid once the surrounding works, including its 8 metre high enclosure wall were under construction. The reference Diagonal also follows a route which was the most level part of the original Giza site.

Imagine a moment over 4500 years ago when Khufu's Surveyor-Priest or even Khufu himself, stuck the end of a staff into the sand at Giza and said, 'We begin from here!'

Part II
Building the Great Pyramid of Khufu

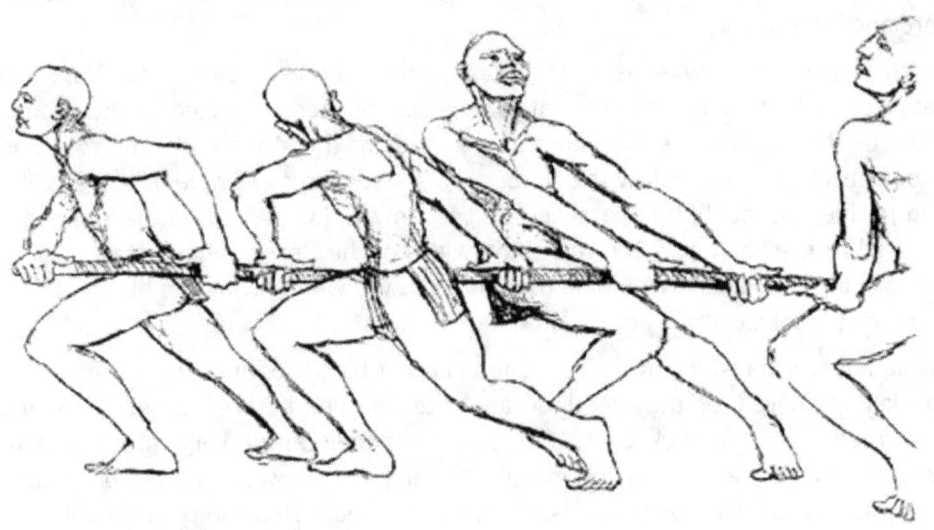

As this is the largest and most accurately finished of all the pyramids, it provides the biggest challenge for describing in detail how the spiral ramp/platform system, the virtual apex method and the virtual centrepoint method, might have been applied in its construction.

It is also against this detail and this pyramid that any alternative methods must be tested and if deficient in any area discarded.

Preparation of the Site and Materials

The stone used in the Great Pyramid is of three types. White limestone from the underground quarries at Tura, on the opposite bank of the Nile, limestone local to the pyramid site and granite from Aswan 600 miles to the south. The following summary is based on detailed studies of Egyptian masonry made by Clarke and Engelbach.

Stone for the Core, Internal Features and Backing Blocks

Limestone on the Giza site was used for the whole of the core and infill for the ramps. It was also used for the majority of the walls, floors and ceilings of the passageways, the Grand Gallery, Queen's Chamber and for backing blocks of the facade.

Stone for the core would be quarried in volume by cutting channels in the rock body and using wedges and crowbars to free the stone in random shapes and weight. In this form they could be used for construction and for the majority be manageable by teams of two to twenty men. Quarry rubble would also be used as core and ramp infill.

Much larger stone was used at critical points above, below and alongside passages and chambers and for lintels. The size of these stones also shows how cautious the builders were when creating voids in the stonework. The roof and floor slabs, above and below the Descending Passage, for example, are over 10RC (5m.) wide, although the passage itself is only 2 RC and other passages are covered with stone up to 2RC thick. The width of the Descending Passage base is partially explained, as it would be used as a slipway when fitting the blocks, which form the passage walls.

These larger stones, including those for backing blocks behind the casing, were specially prepared, as they required accurate finishing before fitting. This stone was freed from the rock by cutting wider channels around all three sides to provide sufficient access for the masons to carry out the work. Each block would be cut to approximate dimensions, before being freed from the bedrock with the use of wooden wedges, expanded with water. Examples of partially freed blocks with these features are still in place at the Giza quarry.

Freeing limestone backing blocks at Giza

Cutting to the North side of Khafre's Pyramid
showing evidence of systematic block removal
Khufu's Pyramid is in the background

Quarrying and Cutting Stone

Casing Blocks

The quarrying of the white limestone Casing blocks used a similar method for freeing them even though they were taken from underground quarries. Channels would be cut vertically to the roof of the cavern around three sides and a horizontal channel cut at roof level. Wedges could then be used to free the blocks from the top down, each being approximately the same dimension.

Pillars of rock were left in place to provide support for the cavern roof and the size of the spaces between indicates that the Egyptians viewed solid stone differently from cut stone, as the distances between were much greater than the spaces they formed in the pyramid. The quarries at Tura were on the opposite side of the river from Giza, meaning each of these blocks had to be transported by barge to the pyramid site.

Limestone

All of the limestone blocks could be cut using copper tools. The total amount of copper required providing and replacing the tools of say 1500 quarrymen and 1000 masons would depend on its wearing properties. As the copper contained impurities, which created a harder material, the wearing rate would have been reduced. If each mason had tools weighing half a kilogram, then over 1 tonne of copper would have been needed to provide the original stock. If the tools were wearing at the rate of half a kilogram every ten days, it follows that a separate workforce had to be mining and smelting a further 1 tonne of copper during the same period.

Granite

The granite blocks used to line the passages and chambers presented a greater challenge. Lumps of Dolerite, a stone harder than granite, appear to have been employed to chip away at the rock, gradually creating a channel around all sides. Fires might also have been lit on top of the rock to make it more friable and freestanding blocks appear to have been used when possible. Whether the lumps of dolerite were hand held or fixed in handles, the shaping process must have been extremely laborious. Unfinished blocks still in place show the channels cut round each side and even under the base of some. Finished blocks could also be freed from the bedrock, using wooden wedges expanded with water. Once free, they could then be taken to the river where crude barges constructed around them, would provide a method of floating them to the pyramid site, exploiting the annual rising waters of the Nile, its direction of current and prevailing wind.

Blocks of stone have been found on the bed of the Nile suggesting that this was not always successful.

The red granite burial coffer placed in the King's Chamber has evidence of both sawing and drilling being used to shape the stone. To carry out this technique, it has been suggested that a slurry containing quartz crystals was used, which slowly abraded the rock, using a copper saw or drill as a guide.

Prefabrication of Blocks for the Façade

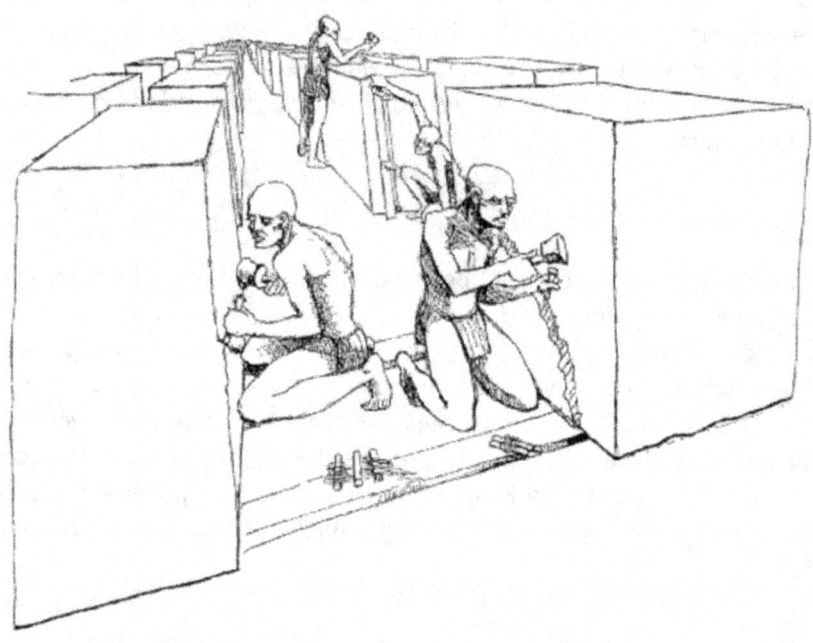

Throughout all the different phases of construction, the casing and backing blocks, which would form the façade, would have been partially prepared.

This aspect of pyramid construction was the least tested and had created problems earlier, therefore special attention would have been paid to these blocks. The accuracy of finishing of the blocks used, particularly the casing blocks together with their size, provides clues to the care the builders were taking with this phase. Clarke and Engelbach made a close study of the few remaining casing blocks in the Great Pyramid, and their conclusion was that they had to be pre-fabricated because of the way they were finished.

They noted that each of the remaining casing blocks has a random angle of rising joint, which meant each adjoining block had to be prepared and finished to the same angle, with each then only fitting to its neighbours.

As this time consuming and accurate work could not be carried out on the pyramid itself, they proposed that all these blocks were prepared on the ground, and in the order they would fit on the pyramid.

The prefabrication of blocks to accurate dimensions was vital to meet the construction timetable, and was only possible because the time and labour force was available, due to the two-stage design.

Backing blocks being quarried near the pyramid site were cut from the rock mass in a controlled way, which required relatively simple finishing.

The casing blocks, transported from the underground quarries at Tura, were finished at the pyramid site so accurately that their touching faces have an average gap of less than 0.5mm in those examples, which remain. The blocks used to form the base course were the largest on the pyramid having a height of 1.4m., with courses above diminishing in height, although with periodic increases in thickness, creating an average course height of 0.7m.

It should be noted that only nine courses of the façade are greater than one metre in height, with over 100 courses as low as 50 cms. Blocks for these courses would obviously be more easily prepared, transported and fitted.

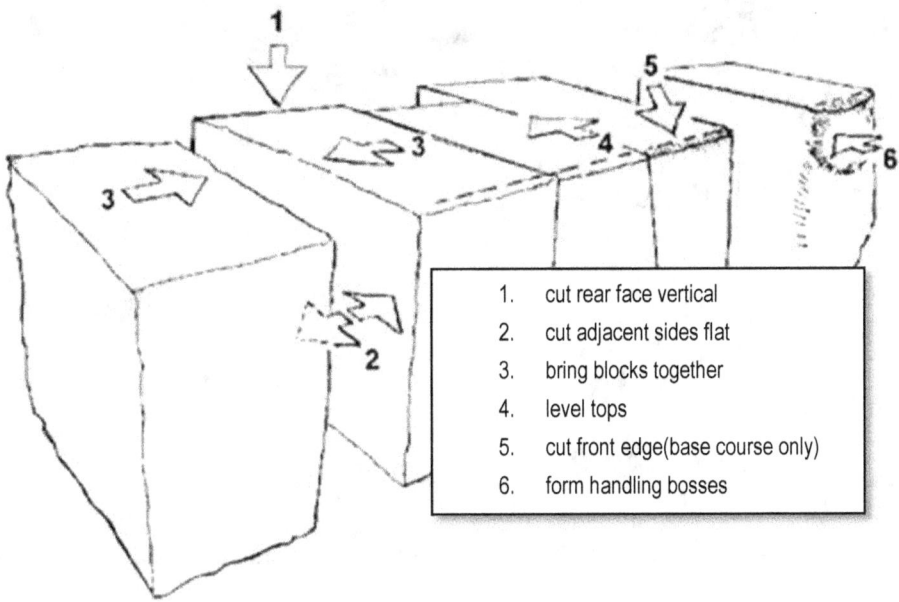

1. cut rear face vertical
2. cut adjacent sides flat
3. bring blocks together
4. level tops
5. cut front edge(base course only)
6. form handling bosses

Preparation of casing blocks

Casing and backing blocks were prepared in different ways, but each would be regarded as being upside down after arriving from the quarries in a rough form. Casing blocks of white limestone of similar height would be placed alongside each other, with the deepest blocks placed in the centre of the row. Evidence for this can be seen in the pyramid today as the faces, now without casing blocks do not run straight and appear slightly concave. The centre of the base for example, indicates that the casing blocks placed here had a depth almost 1m greater than those near the corners.

The preparation of all rough blocks would begin by first cutting their rear face flat and vertical. The sides of each block would then be cut flat and near parallel to its immediate neighbour, but at the most convenient random rising angle, which was at right angles to the rear face.

Finished blocks could then be dragged together so that these carefully cut sides had the closest possible fit. This would continue until the total length of all the blocks or groups of bocks was close to the length necessary to complete one side. The top of the complete line of casing blocks could then be cut flat and level and a chalk coated stringline snapped onto the front edge of each group.

Levelling and marking the underside of base course Casing blocks

The base course only would have had this line clearly marked by having stone in front of it removed, to provide a visual reference for when these blocks were rotated and laid on the pavement layer.

Casing blocks for courses above the base would have had some stone removed outside this line, in order to form handling bosses, leaving only the underside of all blocks unfinished. Four of the largest blocks would have been reserved for the ends of completed rows, as these would later become the corner blocks. They would have only two adjacent sides finished, each cut flat and vertical in order to match the sides of their two neighbouring blocks when they met at right angles to complete two of the pyramid sides.

In order to ensure that rising joints did not overlap the rising joints of the course below, spacer blocks of different widths could have been prepared, which would allow their positions to be modified if required. A more precise, but time consuming method of preventing overlapping rising joints would have been to note the width of each individual block in a complete course.

It would also have been possible to calculate the approximate depth of backing and casing blocks required to complete each course, by measuring the distance from the pyramid edge to the reference core outer wall.

The backing blocks differed from the casing blocks as they were finished with all their sides vertical. Their widths and depths would be noted so they would fit either singly or in multiples, behind casing blocks of similar widths. Their tops were cut flat with their undersides left unfinished. At the time they were to be installed, each block would be inverted onto a sledge and taken in the same positional order to the pyramid. The backing blocks would be delivered first and begin with those placed against the outer walls of the reference core. The block fitting order, at least for the lower courses, would be outwards from the centre of the core sides and forwards towards the face edges, with the final blocks placed being the casing corner blocks of each face.

Block rotation

The rotation of blocks could have been achieved by placing a cradle with a curved base alongside one of the sides of the block and then levering from underneath until the block overbalanced, directly onto the sledge. Models of this type of cradle have been found and its use would not only have helped in rotation, but would also have protected the carefully cut sides of the casing blocks during rotation.

Progressively levering and packing the underside would have aided rotation of the most massive blocks, until the block overbalanced onto a sand bed under which a dragging sledge could have been partially buried.

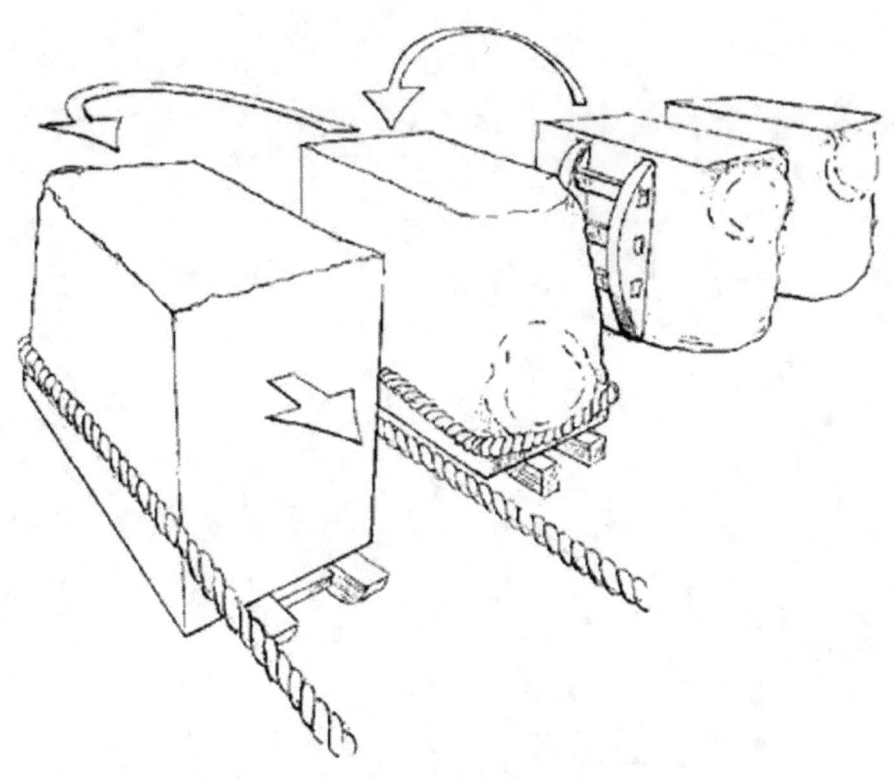

Preparing Façade blocks for delivery

In order for the casing blocks to maintain the same order after rotation they would have to be turned horizontally through 180 degrees, so that the same side, with its unique angle of rising joint, would fit again to the same side of its related neighbour.

Levelling and marking the Base

The base of the Great Pyramid covers an area of over 50000 square metres. However, by exploiting the virtual centrepoint method it would only be necessary to clear the perimeter on which the sides and corner positions of the base square would be placed. The majority of the base could therefore be left untouched and in fact part of the original bedrock rising well above the base is visible today at the partly dismantled northeast corner.

When the surveyors arrived on the empty Giza plateau they chose an area towards its NE corner with the highest visible point and relatively level terrain as the site for Khufu's pyramid. Paradoxically as the pyramid eventually surrounded this mound, another part of the site became the highest.

The central area of the plateau had a steep slope and might have been deliberately avoided by Khufu in order for his project to proceed without delay as it would have been logical to remove material from this area during the construction of his pyramid and use it to add to its core. This part of the site would be progressively levelled and become available for a later pyramid.

The surveyors had to prepare the base square for a pyramid, which would have a centrepoint-corner dimension of 311RC. Their first action would be to form over the uneven ground a square with had a known dimension from its corners to a virtual centrepoint.

Two options are available. A square with sides of 437RC (99X3+70X2=437) with corners 309RC (420+198=618÷2) from the virtual centrepoint, or a square with sides of 495RC (99X5) with corners 350RC from the virtual centrepoint. It appears the latter was chosen, as this larger square later became the base line of the 8 metre tall boundary wall, which enclosed the pyramid and was connected to the pyramid base itself by a level pavement.

Great Pyramid NE corner

Bedrock on the west side of the Great Pyramid

A line of length 495RC and aligned to an eastern horizon marker corresponding to the rising position of beta/delta Scorpii, would have been excavated until level throughout its length. Adding water to the channel would indicate a level line, to which the rest of the base perimeter could be cut. Right angles and partial diagonals could be formed at the ends of this line using the 70-70-99 method to form each. If space was restricted a 10RC-10RC-99 palms right triangle would have been an adequate substitute. Extending the right angles at each end, also through level channels and to lengths of 495RC, would provide two other sides and the corner positions for the fourth side of the square. Any errors in squareness could have been corrected by repeating the measurements and geometry in the opposite direction and confirming that the fourth side also aligned directly to the eastern horizon marker. The bedrock of the perimeter inside this square could then be provisionally levelled to a line just inside where the pyramid pavement layer blocks would be placed.

One final task would be to identify the exact location of a rectangle which would become the entry point of the Descending Passage as it was cut through the bedrock and would also determine the direction of the passage above ground and from this the position of the pyramid entrance in the façade.

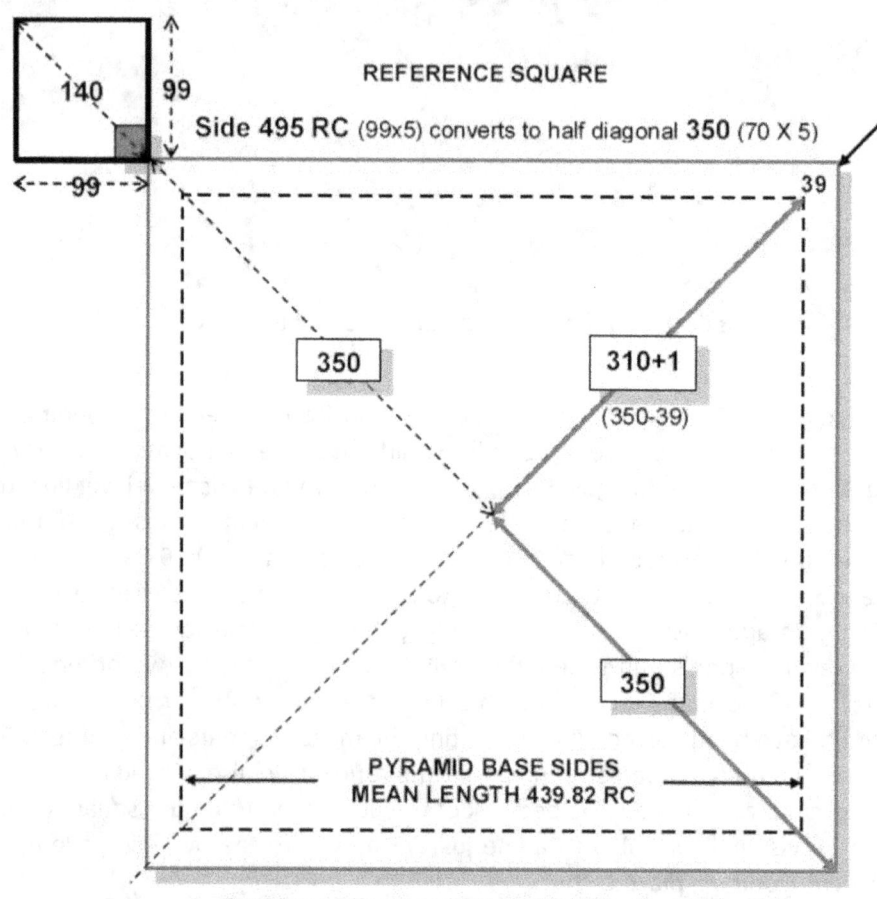

The Great Pyramid Reference Square

The Pavement Layer

Marking out the pyramid base corner positions

The base on which this pyramid sits is of white limestone, laid directly onto the bedrock and it is the upper surface of these blocks, which provide the base for the first course of pyramid casing blocks, being prepared almost perfectly level, with an error of less than 15mm. at only one corner.

The section of pavement which supports the base course of the pyramid has its outer edge cut to a line taken from corner positions, which are 1RC outside the pyramid base corners and are therefore 312RC from the virtual centrepoint. This dimension can be found exactly by measuring 38RC inside the partial diagonals of the original reference square which are known to be 350RC from the virtual centrepoint. Pavement blocks for this section would have their sides prepared before placement so that they fitted exactly to their neighbours. They would be placed with their uncut front edges proud of the line running from the marked corner positions.

When completed, the top of the pavement could be levelled using straightedges and A-frame levels as guides and water added between mud-brick dams to identify high points, which would be progressively removed until a level surface resulted. This is an effective method of achieving the accuracy of levelling found of 1 in 7500 (0.013%). The outer front edge of all the blocks of the pavement layer would then be clearly marked with a line defined by a stringline running between adjacent corners which are 312RC from the virtual centrepoint and therefore parallel with what will become the pyramid base square.

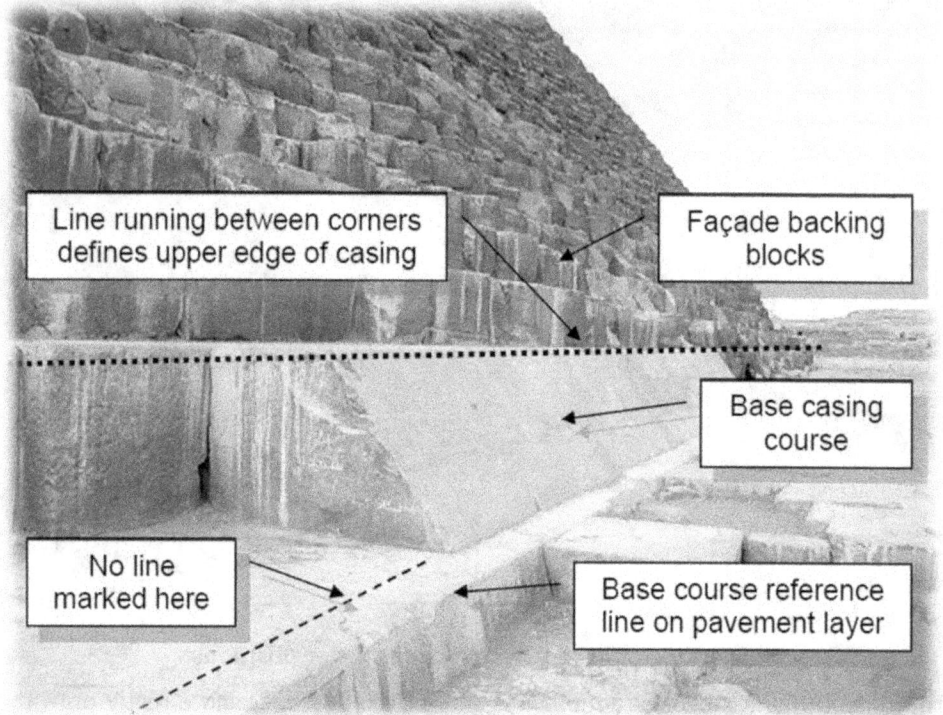

Casing blocks at the Great Pyramid Base

The pyramid base casing course would be laid parallel to and a consistent dimension inside this line, it being the primary guide for ensuring the complete line of casing blocks was straight.

Using the pavement layer front edge as a template provides a practical means for ensuring that the base course follows a straight line. It would have been difficult to use a stringline exactly on the base line itself, as its position would be disturbed each time a block is placed. A stringline is normally used in construction to accurately confirm only the straightness of an upper edge. A base line reference cut into the pavement would also have presented difficulties in its use as it would have been obscured by lubricant or mortar as the final positioning of a casing block with its finely cut base front edge took place.

No line has been found on the pavement layer at the pyramid base edge, confirming that an alternative method must have been employed as a straight line reference.

The Great Pyramid has sockets at each corner, which extended beyond the base edges of the pyramid. These sockets provided two options in their use. If oversize corner blocks had been placed in them, they could be pushed into position to complete the corner after the base course was laid and being oversized would allow a margin of error. Their extra depth would also provide a means for firmly anchoring them to the bedrock base. Alternatively the sockets could have been excavated to take thicker pavement blocks at the corners to ensure they were robust enough to carry the greater weight of the corner blocks above.

Any corner markers which might have been used to secure the stringline when cutting the front edge of the pavement would have been destroyed as these sockets were excavated. However holes are found 10RC outside the northeast corner which could have been used as a reference to replace the corner position later when the pyramid corner edge reference line was being fixed to the base. They might also have been used to provide a right angle here by taking a diagonal of length 99 palms to connect them.

It should be noted that fitting the corner blocks last in a course might not have been the case in other pyramids or even those above the base course in the Great Pyramid. Casing courses in other pyramids show atypical narrow blocks quite close to the centres of their sides. Their appearance here might suggest that these courses were laid from each corner towards the centre, with the narrow blocks sized to complete the final space left between them.

Casing at the base uncovered in 1837

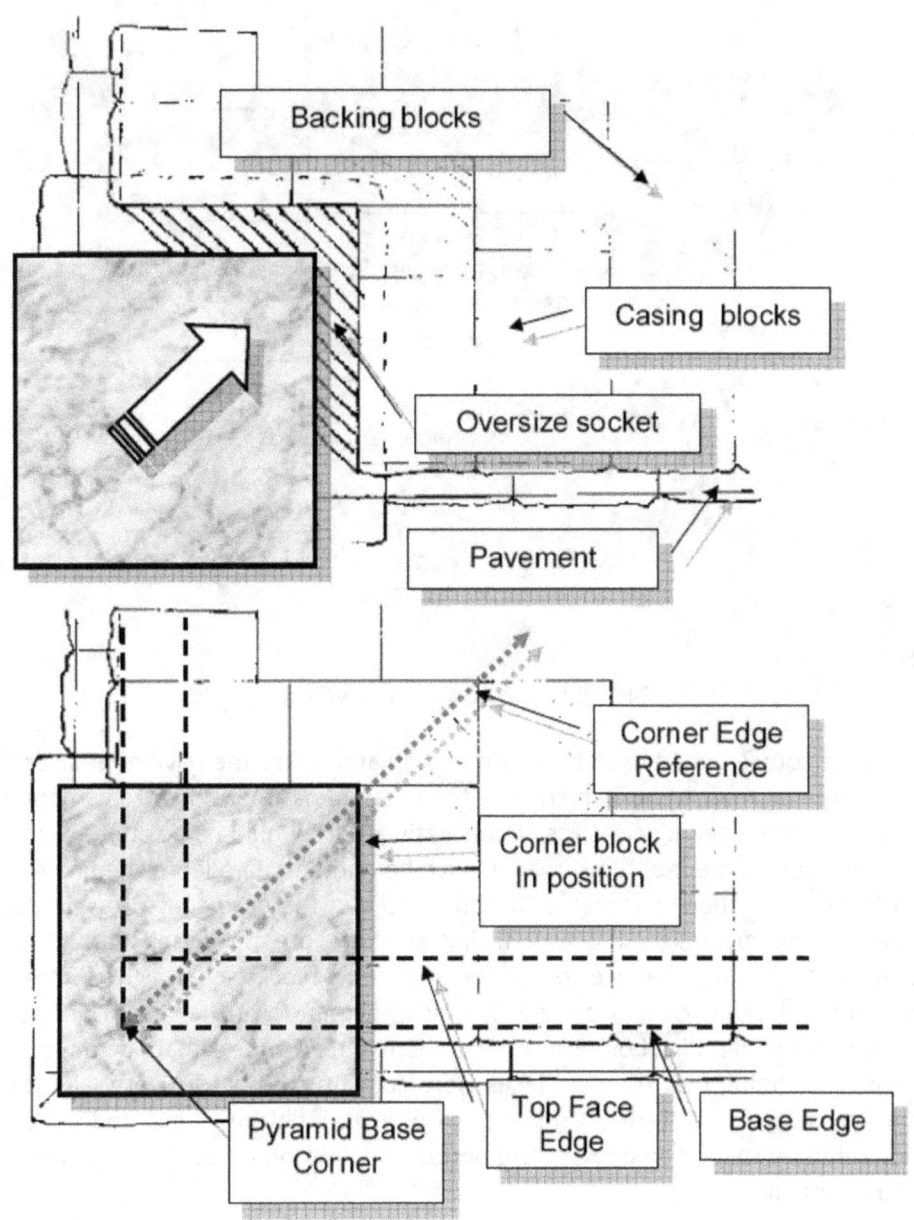

Fitting Corner blocks for the Great Pyramid

Sockets at each corner provide space for adjusting corner blocks to fit the space left between adjacent sides of the casing

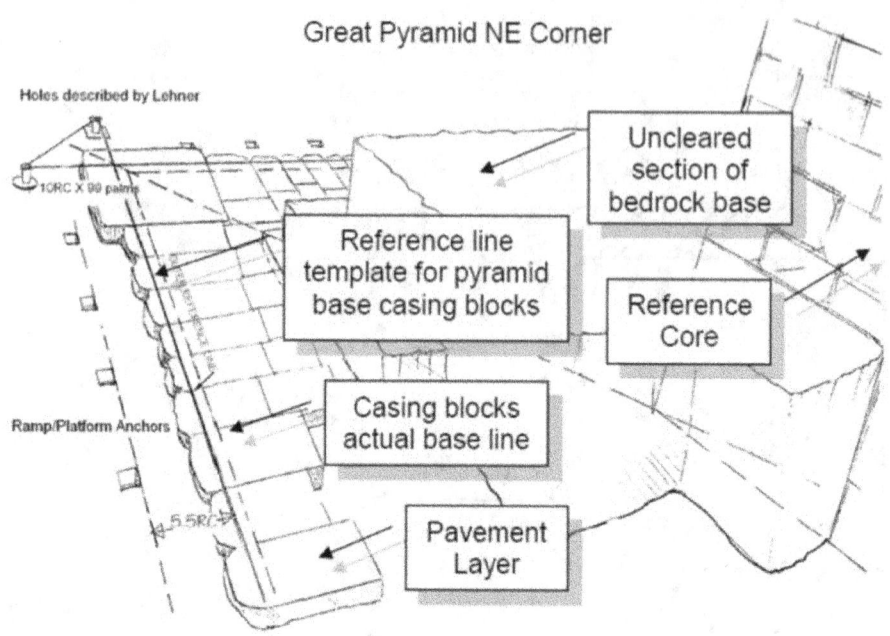

Fitting the Pyramid Base Course

The base course would have been laid immediately after the pavement layer was completed in order to permanently fix the pyramid perimeter before work on the reference core began. Each base course casing block would be positioned with its carefully cut base edge, the same distance behind and parallel to the line cut on this layer. It has been suggested that the whole of the base course casing blocks were completely finished before fitting as there are no chisel marks on the pavement layer and the faces of the few remaining blocks show slight differences in flatness. However other pyramids, which have parts of their faces unfinished show only the handling bosses formed in them and an approximate pyramid slope angle, confirming that the final faces were cut after the blocks were laid. It was not necessary to finish the casing block front faces before fitting, and would be illogical to do so as they do not form part of the method for controlling the shape of the pyramid.

It would not only make moving, fitting and alignment more difficult, but blocks would also be more easily damaged, causing serious delays if they had to be replaced.

The slight differences in flatness of the base course could have been caused by difficulties in placing straightedges accurately when the pyramid faces were being cut on completion of the whole pyramid. Not only were these blocks extremely large, but their base edges connected directly with the pavement layer restricting the use of this tool. As the base edge itself was already completed the masons would have easily prevented their tools from damaging the pavement layer below

By investigating what had to be achieved in fitting the base course, the tallest of the pyramid, gives clues to the fitting of all blocks.

The largest of the remaining casing blocks weighs almost 20 tonnes. It would have taken a team of 200 men to drag this block on level ground, but on reaching the pyramid there would be insufficient space for 200 men to continue pulling this block, as it would have required 50 metres in front of them. Traces of a crude mortar have been found between the faces of the blocks, and its composition resembles a lime mortar or plaster. It has no function as cement as the blocks fit so closely, but it might represent the remains of a lubricant used to allow the blocks to be positioned more easily. Modern masons use a similar system today to position blocks of stone with a lubricant they call butter, which allows easy small final movement to be made when adjusting a block of stone. The lubricant used by the Egyptians might have been this crude mortar, which if it had a cellulose size or ammonia added, would have had increased its effect as a lubricant and retarded its setting time. Lubrication would be essential at this time in order to complete the final fitting speedily and with the minimum of effort.

It has been shown that only 6 men can move a 2 tonne block, when the base on which the block moves is smooth and properly lubricated

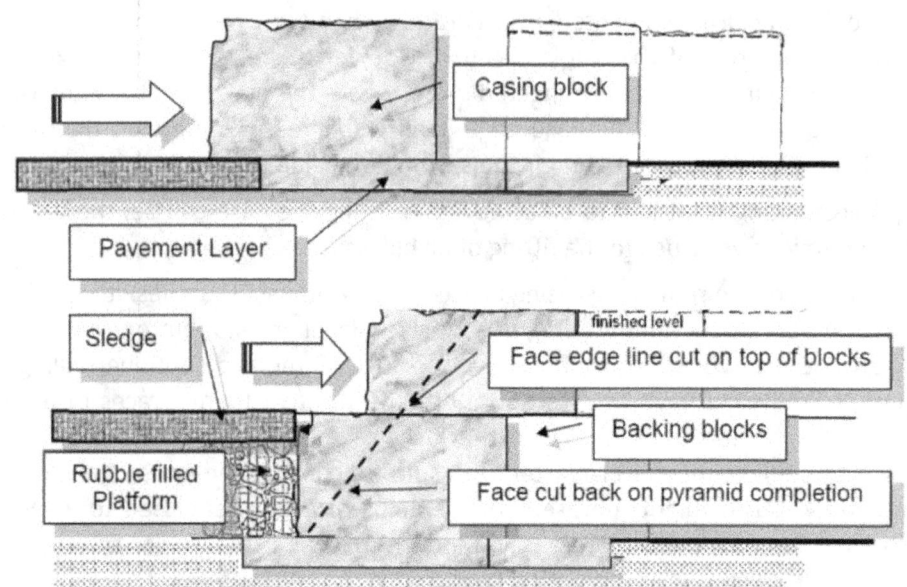

Fitting the base and first course of the Façade

The so called levering bosses found on other unfinished casing blocks could instead have been bosses to retain rams which would have been used to push the blocks into their final positions, as it was horizontal movement not lifting which was required. The pavement layer, if level with the top of the material sledges, would allow blocks to be pushed evenly on to this smooth level base.

The fitting teams, placing both casing and backing blocks could have used both pushing and lubricant to remove blocks from their sledges, onto the pavement layer and into their final position. The most effective way of achieving this would be to place one side next to its lubricated neighbour and then to push backwards, using the adjacent block as a guide. Its final position would be when its base edge was parallel to and equidistant from the pavement edge and its rear set against the vertically cut front face of a backing block immediately behind.

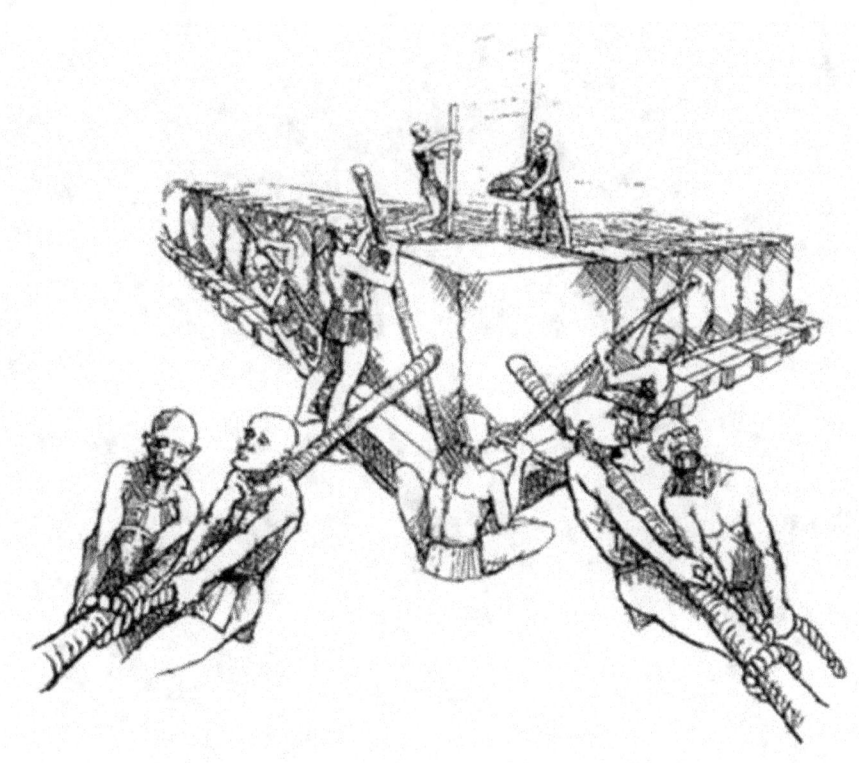

Pushing a corner block into its final position

When the four corner blocks were in place the base course would have the front faces and top of each block left unfinished. It would then be covered with loose material to provide protection when construction was directed to building the reference core to its first reference height.

The Reference Core

The first objective after marking the base square, laying the pavement layer and the first course of the façade, would be to build the core structure to one of the known reference heights. The first height would have to be greater than the central mound in order to provide a level surface on which the centrepoint could be found, by using diagonals running from corner to corner.

In the Great Pyramid the minimum reference height would have to be at least 19RC above the base as the un-cleared section is at least 7m. high. Other considerations might also have affected this decision.

The Great Pyramid has many interesting and complex features within the core and each had to be installed at a pace which matched the overall construction timetable. Some of the blocks from which these are formed, are the largest and heaviest in the pyramid, with weights possibly up to 50 tonnes.

Access to the core during construction to the first reference height only would have been from a single perpendicular ramp attached to one face and it would have been logical to take all the heaviest blocks as high as possible using this gradually sloping, wide and straight ramp. If one of the lower reference heights had been chosen then all these large blocks would have had to be taken up the much narrower spiral ramp attached later to the façade.

The height at which the Queen's Chamber was placed also gives a clue to the first reference height chosen as it was built on top of the core when it had reached a reference height of 37RC. If the pyramid was constructed in alternating stages of core then façade, this chamber could have been assembled here during the time the façade was being added to the same height. The entrance to the pyramid might have been a complicated structure with a hinged swivelling door, now missing. However, the upper section of the Descending Passage visible at the entrance is at a height of 35RC and is surmounted by huge blocks forming a vaulted arch. It would have been ideal if the installation of these features had taken place when construction had moved back to the core as it was taken to its next reference height at 82RC.

Other large blocks, to be used later to form the King's Chamber at the next reference height, could also have been delivered and parked at this level.

As work continued on the core around the Queen's Chamber these blocks could have been taken progressively higher, using internal ramps built on the core itself to take them to subsequent heights.

For these reasons it is reasonable to assume that the core in this pyramid would have been built to a first Reference Height of 37RC

The core structure in its simplest form is a tower surrounded by buttress walls, at regular intervals to its perimeter. This design creates an ideal working area for large numbers of masons. As the core rises it must stay within the boundary of where the pyramid outer corners will be and it would have been reasonable to prepare for this in the dimensions of its base and the width and number of buttress walls.

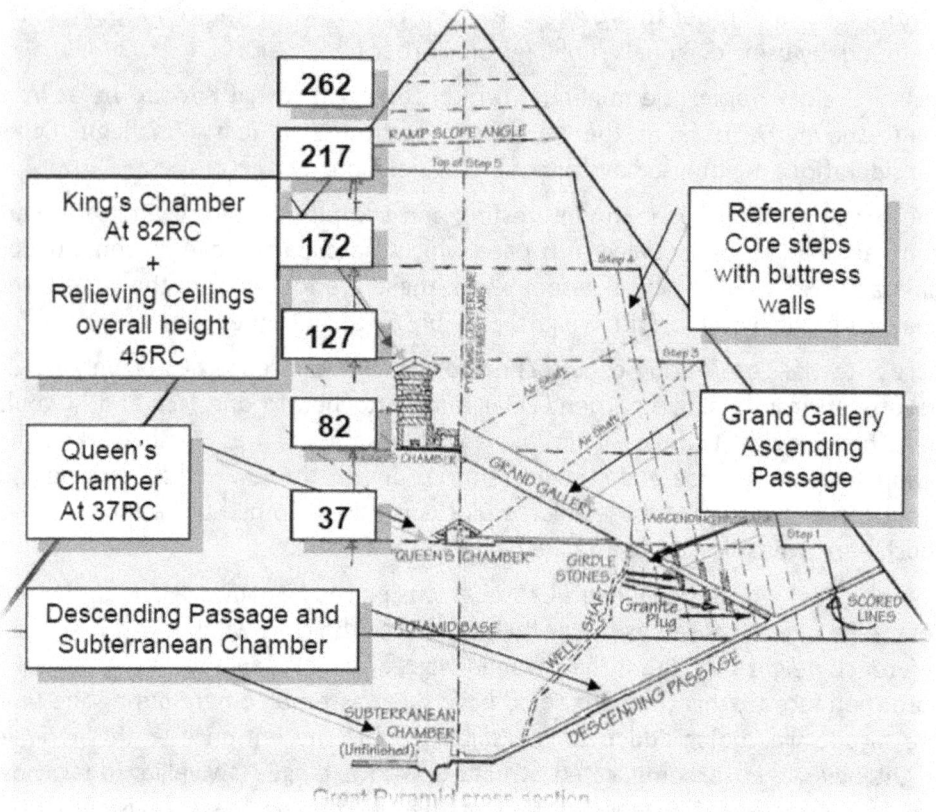

Reference Core and Internal Features

A 5 or 6 stepped core with a base centrepoint to corner diagonal of 252RC, buttresses wall corners at 14RC intervals, each corner sloping at 3:1 ratio and step heights of 45RC meets this condition, although other configurations are possible, including one based on reference heights. The steps can be formed simply, by eliminating two outer buttresses at each new step height. If the buttress wall corners were at intervals of 14RC along the diagonal, over 5000 metres of parallel walls would be under construction at the base. Each corner of the tower and buttress walls would be constructed first to provide a straight-line reference for the wall faces. A slope of 3:1 at each corner creates a side face slope close to 78 degrees and this angle of slope has been found in other stepped core structures, which have their corners exposed.

The space between their walls has also been measured at +-10RC, which would confirm a diagonal interval of 14RC (or 99 palms), between the buttress wall corners.

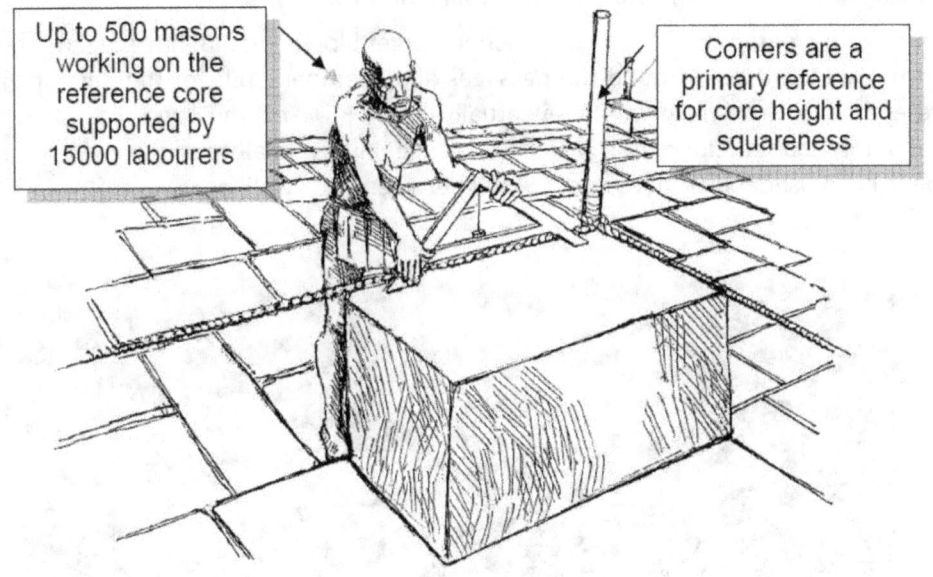

Buttress Wall corner construction on the reference Core

Over 500 masons could be working individually or in teams, to complete less than 1 sq. m. of buttress wall in-filled behind to a depth of 10RC (5.5m.), each day. Each inner wall would be built slightly in advance of those outside to allow the accurate control of the corner slopes, with stringlines used to keep the faces straight.

However, as it is the corner positions and their height, which provide the vital dimensions for controlling the shape of the façade, it only these which would have to be monitored routinely and carefully to ensure the structure stayed square and of known height.

Wall construction today still follows exactly the same procedure, with corners accurately prepared first and the face sections subsequently in-filled to lines taken between them. The 'unfinished pyramid' at Saqqara for the Pharoah Sekhemket for example, has a core constructed of randomly shaped stones, but to accurate corner slopes of 3:1. Random stone construction is indicated in the

Great Pyramid also, as the internal structure shows some settlement, while the façade built with closely fitting cut stone, has none. The core structure stonework is different in the Great Pyramid, as the blocks used were more regularly shaped due to the way they were quarried, but still random in a general sense and core drilling has reported limestone, mortar, sand and voids.

During the early stages of construction large blocks would have been used throughout and also around the passages and chambers, but for the bulk of the reference core, which would eventually total 85% of the pyramid volume, randomly shaped blocks, the majority manageable by small teams of men would have been used, particularly for the infill between each wall face.

Core structure – Sekhemket's Pyramid

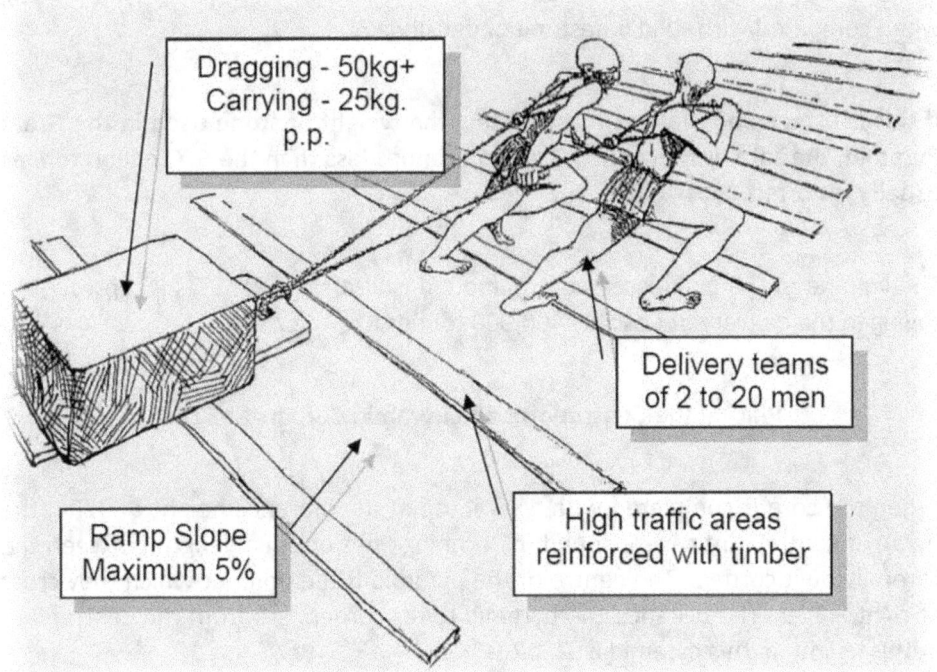

Delivering stone to the Core

Gangs of labourers arriving on the pyramid core would be directed by strategically placed foremen, to those masons who required material. On receiving material the mason would select stone for either face or corner construction, or infill, and in the latter case the stone could have been placed directly by the labourers themselves. The necessity for a process such as this, allowing stone to be taken in one movement from the quarry to its final position, is that it reduces double handling and becoming 'muck-bound' as a consequence; the scourge of all large civil engineering projects. The elimination of double handling is a fundamental part of the planning of all such projects, as it is wasteful of both space and effort and especially so in solid pyramids, when the volumes of material are massive and the working areas limited.

The use of random stone is effective in solid structures as it is both rapid in preparation and structurally sound, when certain conditions are met. Mortar or rubble can be used to infill the spaces between blocks to improve stability and rising joints should not overlap. Egyptian pyramid design also ensured that the whole of the core structure was encased and buttressed by a more carefully constructed façade.

There is a significantly different density for this kind of stonework, typically 1.5, when compared with solid limestone of density 2.6.

If this is used as one factor in calculating the weight of stone used in the Great Pyramid, the total is more than 2 million tonnes less than the 6.7 million tonnes usually reported.

The volume of the un-cleared central mound is also significant, as it represents a saving in the delivery of possibly 400,000 tonnes of stone.

Placing the Centrepoint and Pyramid Corners at 37RC

When the core is completed with a level top at its reference height of 37RC, the pyramid centrepoint can be found, by running lines of equal length between the carefully built corners. The centre of the pyramid is the point at which they cross at right angles. The distance the pyramid outer corners are from the centrepoint at this height, in this pyramid is: 270RC.

$$(280 - 37 = 243 \div 9 = 27 \times 10 = 270)$$

In order to place them in this position, horizontal markers in line with the diagonals could be extended to this dimension beyond the core.

Lehner describes how one of the 'Queen's' pyramids of the Great Pyramid has a series of holes on top of a step of its core and in line with its diagonal. These might be the remains of anchor points used to secure an extended diagonal of this type.

Stringlines connecting these upper corner markers to the base corners will define the four corner edges of the pyramid to this height, each aligned with the virtual apex, exactly 280 Royal Cubits above the base.

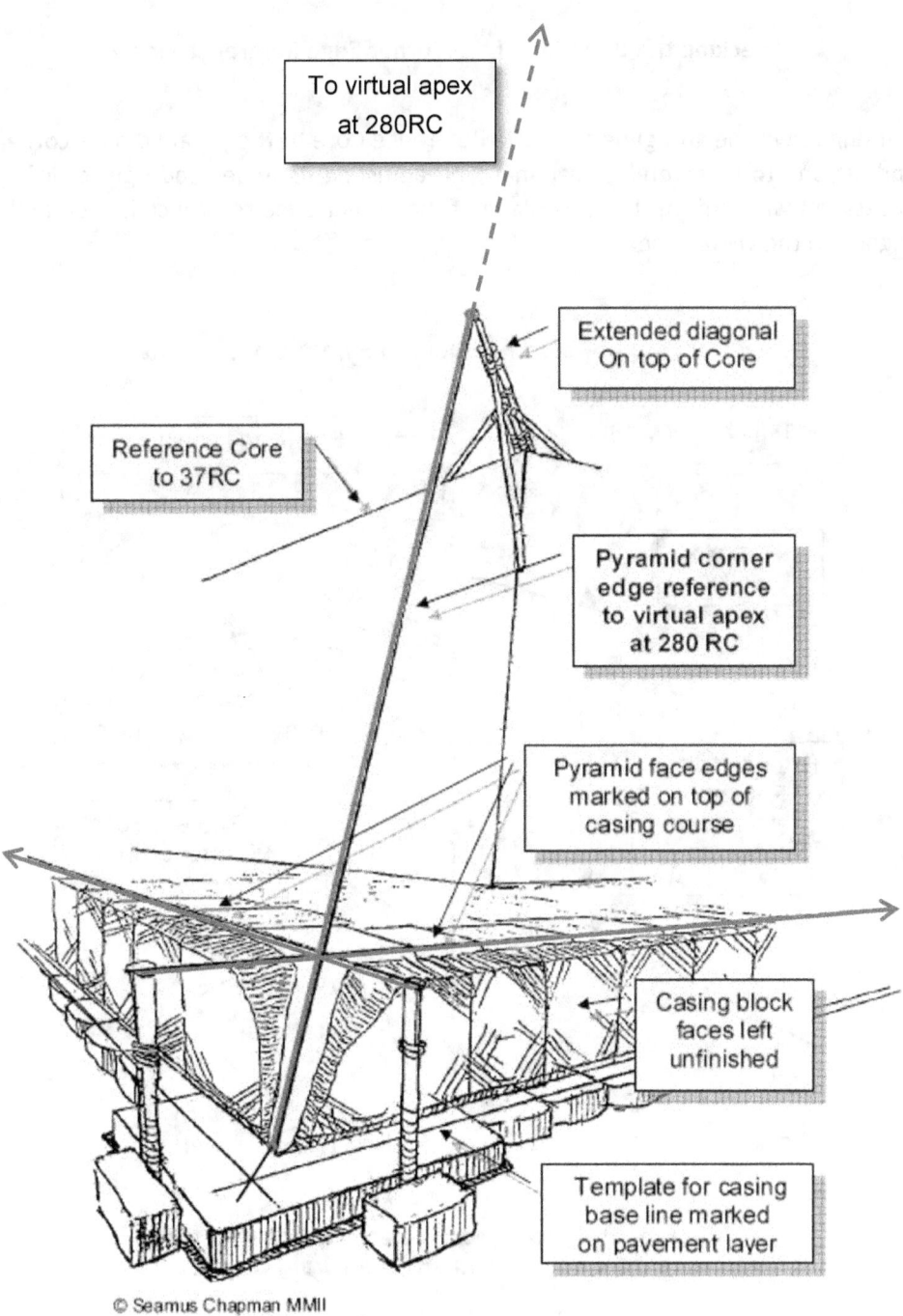

The pyramid Corner Edge Reference Line

Checking the direction of the Corner Edge Reference Line

Sighting down the stringline from the Reference Core to the pyramid base corner and beyond to its extended diagonal (and adjusting its upper end right or left if necessary) will confirm that one plane of the corner edge reference is accurately aligned to the virtual apex.

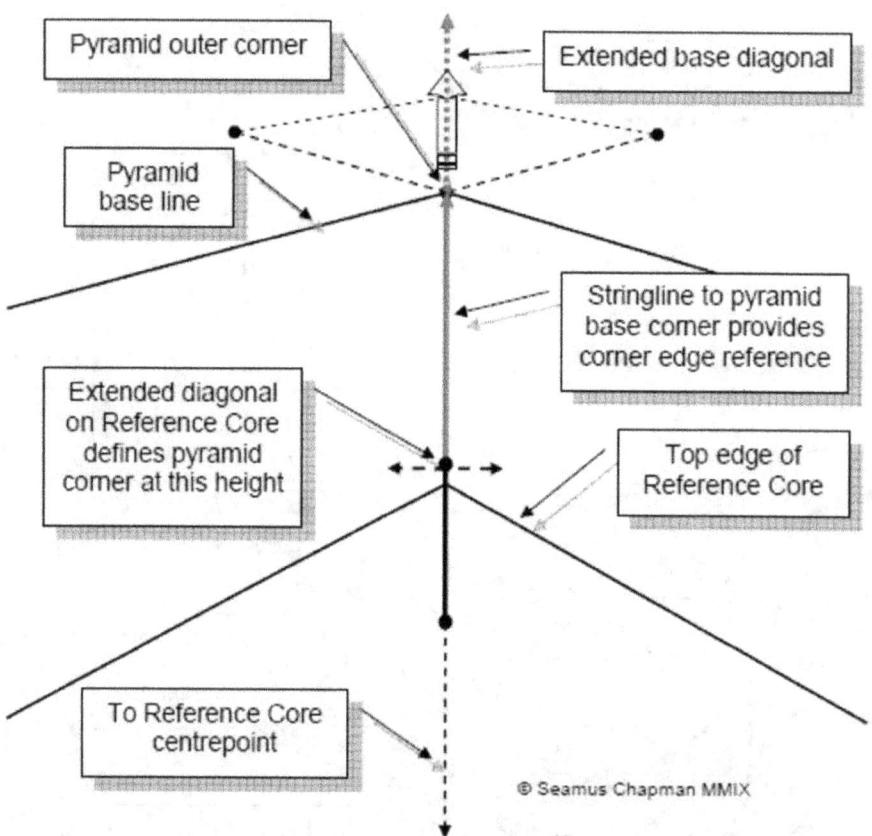

Aligning the corner edge reference line with the Reference Core centrepoint and the pyramid base corner

Taking the most massive blocks to the top of the Core at 37RC

A final task before the workforce moved from the core to begin the installation of the façade would be to deliver all the massive blocks for the Queen's Chamber - Grand Gallery Complex and those for the King's Chamber – Relieving Ceilings to the reference height of 37RC, using the broad perpendicular ramp.

With the time available during façade installation, easy access and plenty of space on top of the core, a unique opportunity was presented. Although animals would generally be a hindrance in pyramid construction they could have been profitably deployed here to pull the heaviest blocks to 37RC and also be available for block movement on top of the core as the Queen's Chamber and Grand Gallery were assembled.

Adding the Façade to 37RC

Levelling the Top of the Façade Base Course

The base course which was already in position would now have its top levelled in readiness for marking the corner edge on the corner blocks and from these the crucial pyramid face edges on the top of the blocks forming the sides. Levelling the top of this, the tallest course would have employed 'A' frame levels and straightedges placed between vertical guides which used the pavement layer as a datum. The backing blocks immediately behind would also be cut flat and to the same level using straightedges and A-frame levels taken from the top of the levelled casing blocks.

Marking the Corner edge and Pyramid Face

As the whole course was being levelled, the corner edge could be cut into the corner blocks to match the direction of the stringlines running from the upper reference core markers to the pyramid base outer corners. A stringline fixed between the upper corner edge positions on two adjacent corner blocks would clearly define the pyramid face edge at this height. The complete course of casing blocks would have this line clearly cut into its top. At the base the stringline would be over 200 metres in length but as it rested on top of the course of blocks, its straightness in the horizontal plane would be exact when stretched tight.

Courses above the Base Course

The fitting of casing blocks for courses above the base would be slightly different from the base course as their base front edge was not marked or cut on these blocks before fitting. In each case a casing block would be pushed into place alongside its matching neighbour and checked to ensure that its unfinished front face extended outside the face edge line marked on the course below.

In order to carry out the installation of all bocks it was vital that a platform, level with the top of the previous course and running round the whole structure was in place. Blocks arriving on the platform could then be taken either left or right, around the exterior of the pyramid to their appropriate and pre-determined positions. Levelling the top of the course would begin immediately a block was laid using the same method as the base course to ensure they maintained a consistent height above the course below. The access platform would be raised to their upper edge during the same period.

When all the casing blocks for a side were in place, stringlines running from the upper corner edges of the corner blocks would define the pyramid face edge on the complete course and a permanent line would be cut into their levelled tops. The platforms surrounding courses immediately above the base could be accessed from small feeder ramps, but as the façade grew in height the perpendicular ramp used earlier to construct the core would be modified at its upper end so it connected to each new platform height.

The addition of the façade would continue to be installed in this way, until it became level with the reference core at a height of 37RC.

One of the final elements installed in the façade as it approached 37RC would be the massive blocks forming the arch over the pyramid entrance and the sophisticated mechanism for the camouflaged door. While this work was being completed all the massive blocks for the King's Chamber and Relieving Ceilings delivered earlier to 37RC, could be taken up to 82RC using the ramp and platform system built around the Queen's Chamber–Grand Gallery complex as they were assembled. Once again as the time and space was available, animals could have been deployed to assist in this task.

Construction Elements of the Great Pyramid to a height of 37RC and above

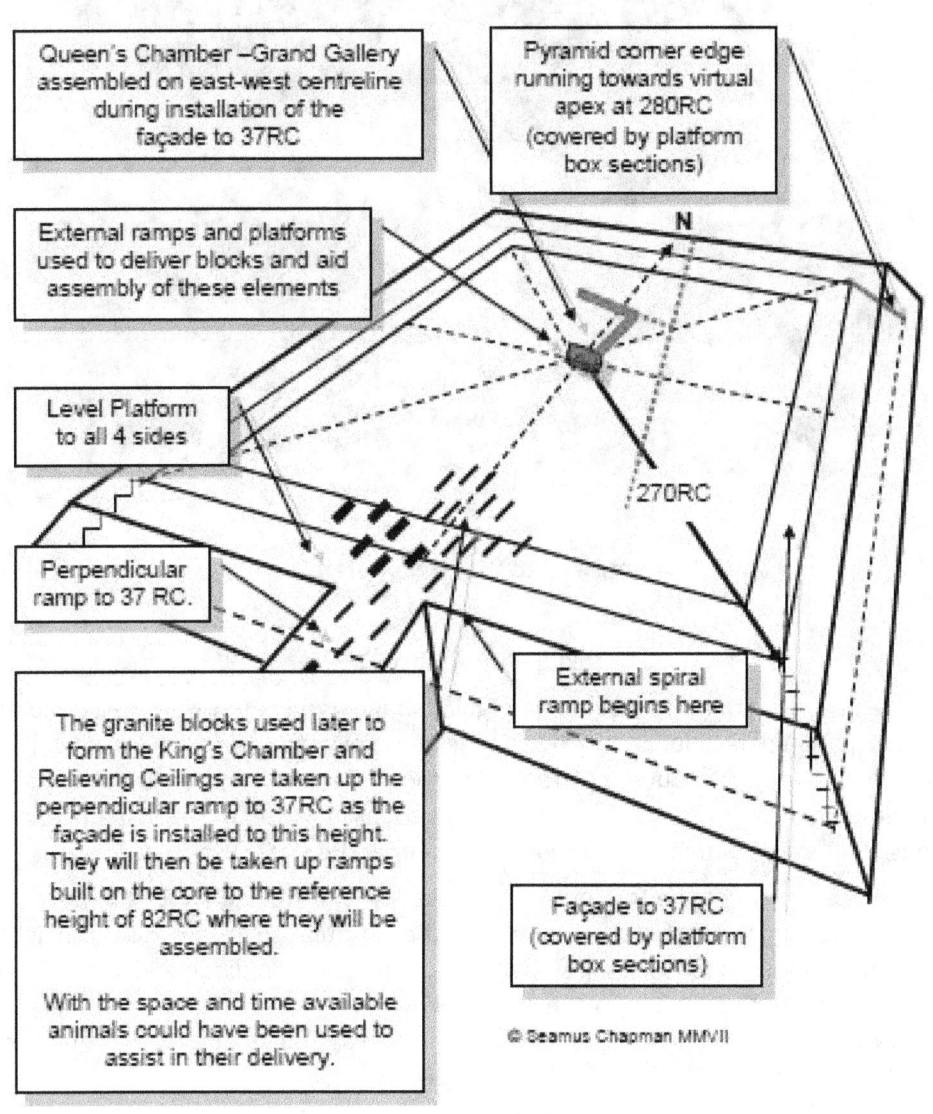

Entrance to the Great Pyramid

Taking the core to 82RC

The reference core must maintain a regular shape and keep within the boundary of the pyramid façade as it grows in height. This can be readily achieved by omitting two of the outer buttresses at the 37RC level and each subsequent increase in height of 45RC. A core of this design with 6 steps maintains a consistent shape within the external dimension of the Great Pyramid.

The space on top of the two outer buttresses would provide an ideal base for a broad core ramp, which would spiral round each outer wall as it rose in height. The massive blocks for the King's Chamber and Relieving Ceilings, which had been delivered earlier to the top of the core at 37RC, would be progressively taken up the ramp used to install the Grand Gallery, to be assembled at the next reference height of 82RC, 45RC above the Queen's Chamber base. All the material for the core would be brought up the perpendicular ramp and pass over the finished façade.

Once the core was completed to 82RC and with a level top, careful measuring would determine the diagonals, centreline and centrepoint and from these the location and length of the extended diagonals based on the known remaining height to the apex of 198RC.

$$(280-82 = 198 \times 10/9 = 220)$$

Once the extended diagonals were in place 220RC from the centrepoint and the pyramid corner edge reference lines were running from these to the corner edges at 37RC, work could begin on installing the façade to take it to this new height.

The same period would be used to install the King's Chamber, offset from the known centreline of the pyramid core, in order to position the granite sarcophagus directly on it.

The only workers coming onto the core as the façade was being installed would be those positioning the King's Chamber blocks or those delivering material for the access platforms and ramps around it.

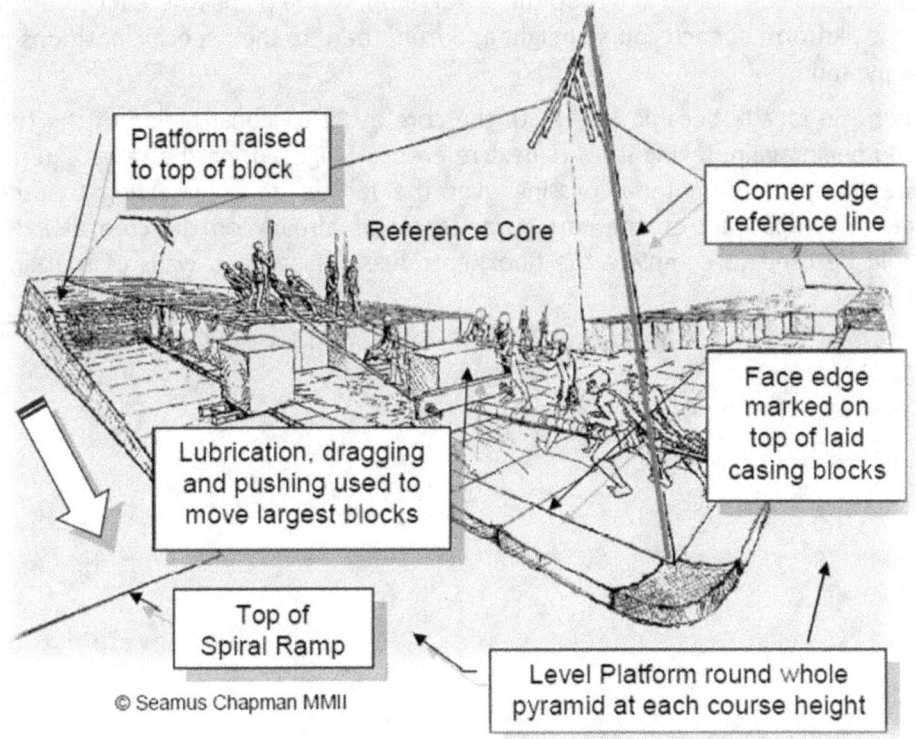

Elements of the construction processes when adding the façade

Taking the façade to 82RC

As the perpendicular ramp only extended to a height of 37RC, a new ramp would be formed at right angles to its upper end and connect to the outer edge of the external platform already in place.

This 2m wide ramp would run parallel with the platform, increasing in length as it is connected to each new platform height, eventually spiralling round the pyramid as it turned corners and rose in height.

If the ramp and platforms maintain the same geometry, after each complete circuit of the pyramid, the base of the ramp will sit directly on the now redundant platform of the previous circuit.

All the prefabricated casing and backing blocks for the façade would be taken up the perpendicular ramp and onto the spiral ramp, to the point where it connected to the platform at each course height and from there to their specific positions on the pyramid.

When the façade became level with the core at 82RC, construction of the core could begin again, taking it to its next reference height, using the same external spiral ramp, with material passing over the façade and connecting to other internal ramps built on the core. Granite blocks already on the core at 82RC, would be used to complete the floor, entrance passage and walls of the Kings Chamber.

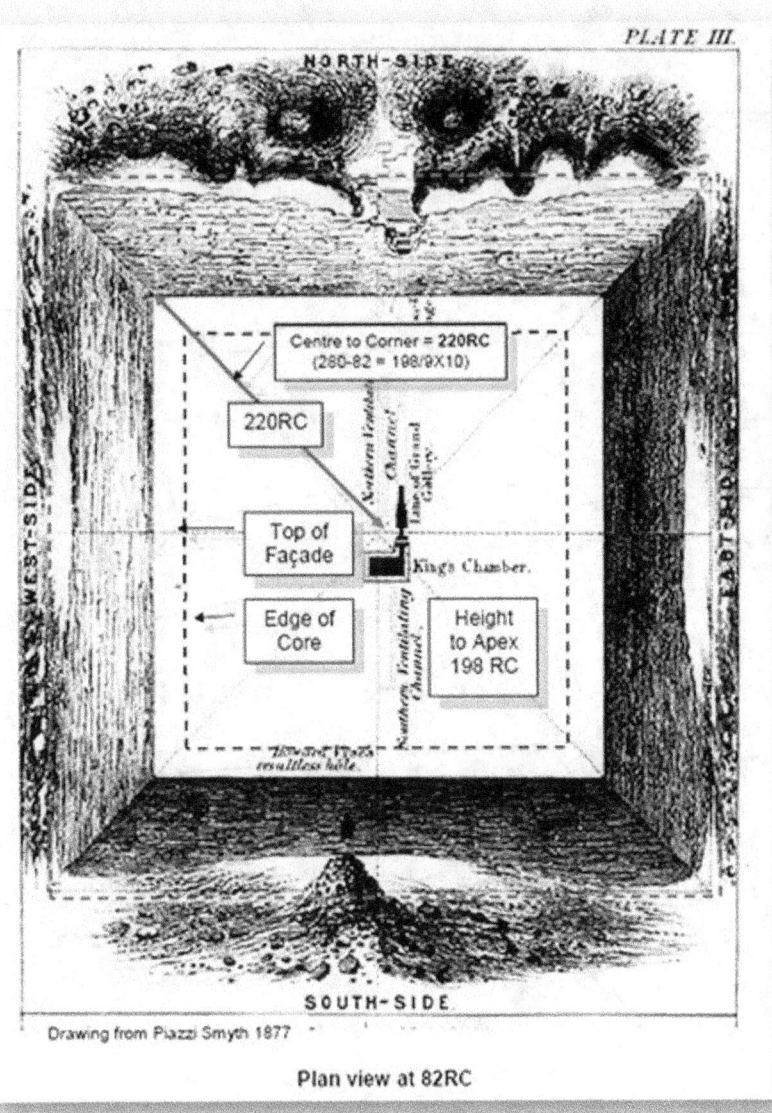

The Great Pyramid at 82RC

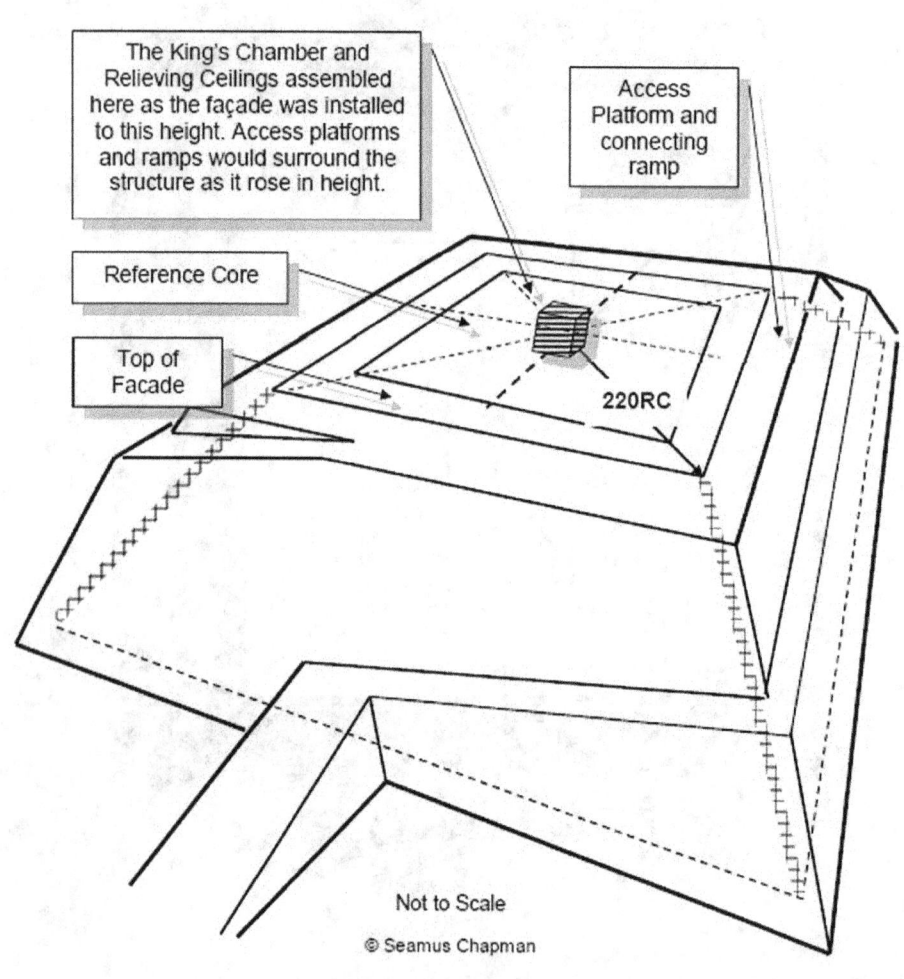

King's Chamber installed as the facade is taken to 82RC

Construction stages from the base to the apex.

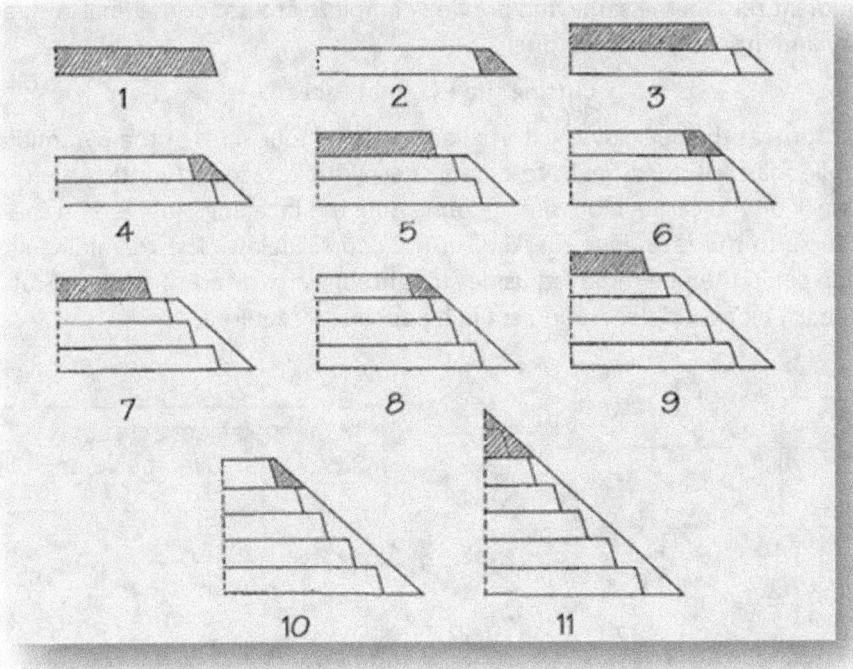

The two phases of alternating construction of core, then façade, are repeated at each new reference height in intervals which are any multiple of 9RC and a known height remaining to the virtual apex divisible by 9.

Both would eventually reach a height 262RC with the pyramid corner edges now 20RC from the centrepoint at this height and each running straight from the base towards the virtual apex 18RC above.

During each phase, the spiral ramp has been available to take materials either onto the core, or to external platforms surrounding the façade. Completing the final 18RC to the apex could have been carried out in one stage, using a height 9RC from the apex to check that the corners were 10RC from the centrepoint. Straightedges extended upwards and in line with the completed corner edges at this height, would provide the final guide for positioning and marking of the rest of the casing and capstone.

At lower levels the ramp would have been rising some 20m on each circuit to maintain the 5% gradient. For the final section to the apex, with shorter circuits, the low ramp gradient would be maintained, by making the pitch of the ramp and platform outer support walls steeper. In this way the ramp and platform widths and gradient remain the same and provide complete access for the final and rapid delivery and fitting of the capstone.

Cutting the Pyramid Faces

The platform at the apex gives external access to all four sides of the pyramid and from this; masons could begin to reveal the pyramid faces by cutting away the extra stock on the casing blocks, first connecting the face edge marked on the top of a course to the face edge marked on the course below. The remaining stone between could then be removed using straightedges to ensure the flatness of the face of each block and decoration could be applied if required.

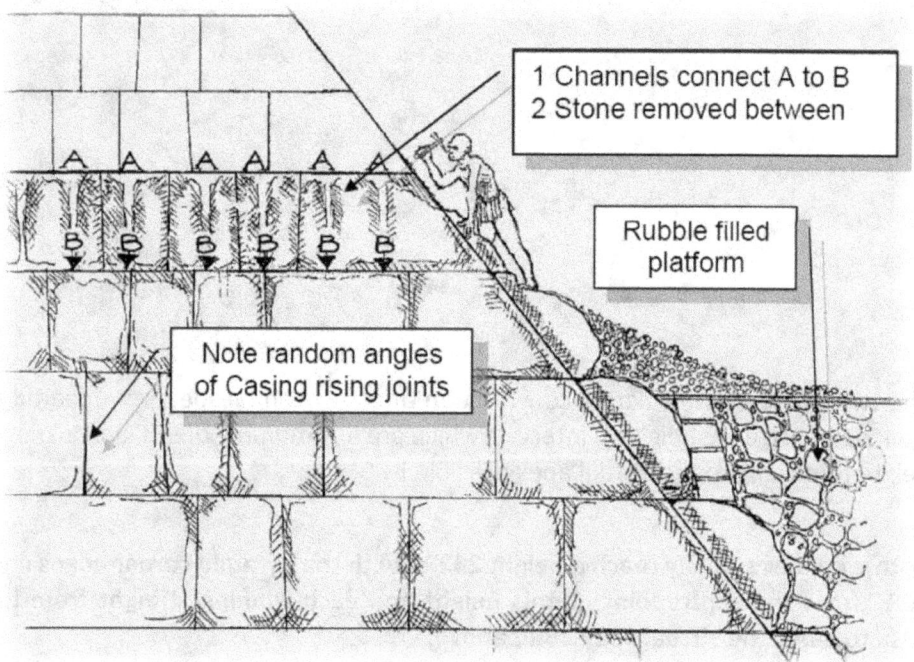

Cutting the pyramid faces to fit the face edge lines inscribed on the top of the casing blocks as they were placed

The platform and ramp would be progressively dismantled, exposing new casing to be cut back and work would continue in this way to the base of the pyramid, with a total surface area of 85000 square metres completed.

The comprehensive access from the platform at every height would allow space for an increasing number of masons as they progressed down each face. If an average of 150 square metres was finished flat each day, then all four faces could have been completed in 600 days.

During the same period an average of 400 cu.m (600 tonnes) of ramp and platform material would have to be removed and dumped each day.

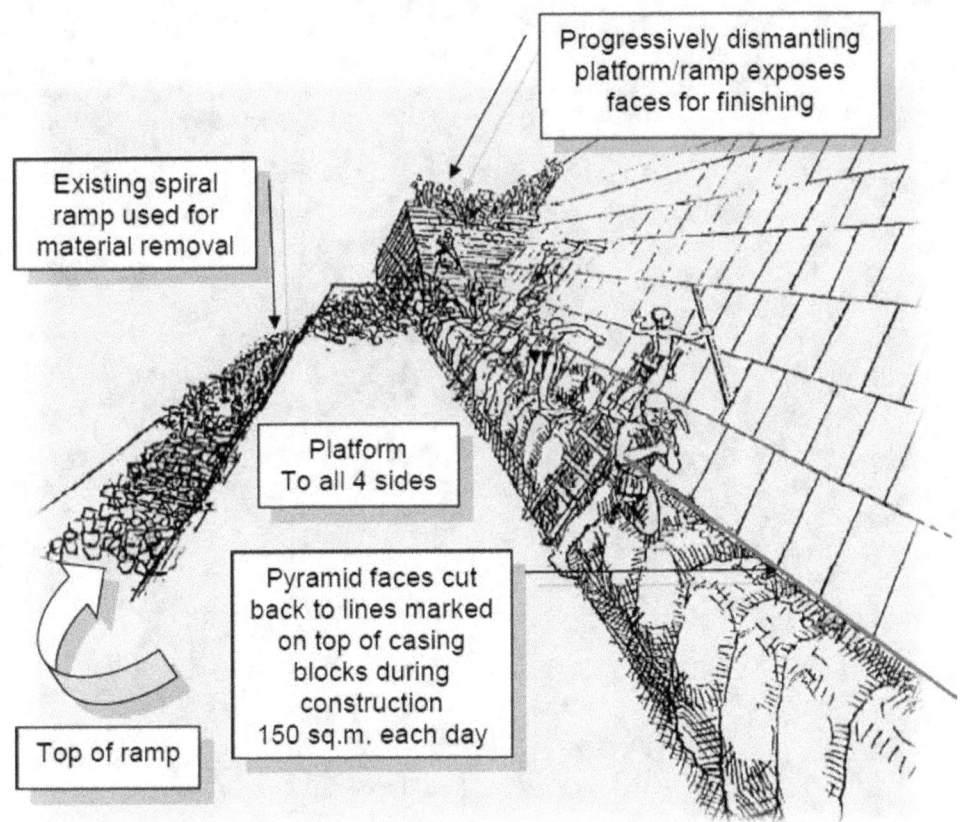

Progressively dismantling the platform/ramp system exposes more casing

When completed the pyramid had four flat faces sloping at an averaged angle of 51º50'40", but with each corner edge running straight to the apex, 280 Royal Cubits vertically above the virtual centrepoint of the base.

The base square had unmeasured and unequal sides with an average length of 439.82 RC.

Internal features of the Great Pyramid

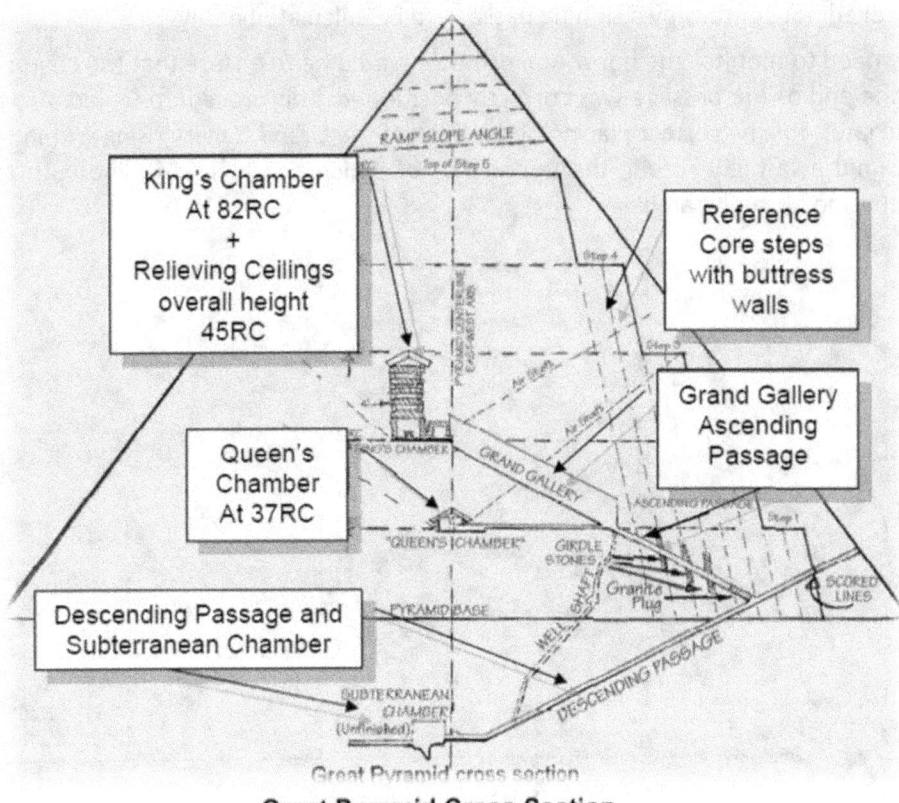

Great Pyramid Cross Section
Looking West

The internal features of the Great Pyramid are described as the Subterranean Chamber, Descending Passage, Well Shaft, Ascending Passage, Queen's Chamber, Grand Gallery and King's Chamber.

The original intention appears to have been to provide a subterranean burial chamber and the work on the passage which led to it, would begin as soon as the site levelling was completed. The dimensions of this passage and its angle of slope, part of which was cut through solid rock, suggest that both sections were planned before the work on the pyramid began. The dimensions and position of the original rectangle which was marked on the base to define the below ground section, creates all the dimensions of both the tunnel and the passage above, including the pyramid entrance which was offset from the centreline 13.88RC to the east.

The Descending Passage forms the hypotenuse of a right triangle with a base which is twice the height, creating the angle found of 26º31'.

The importance of this geometry is that it also allows the distance which is travelled horizontally by a sloping passage to be calculated easily.

The need to monitor the horizontal distance would be to ensure that the chamber at the end of the passage was correctly positioned. This procedure would also be vital later, for the correct placing of the base of the Grand Gallery rising within the pyramid at a higher level, the upper part of which had to end in line with the centreline of the pyramid.

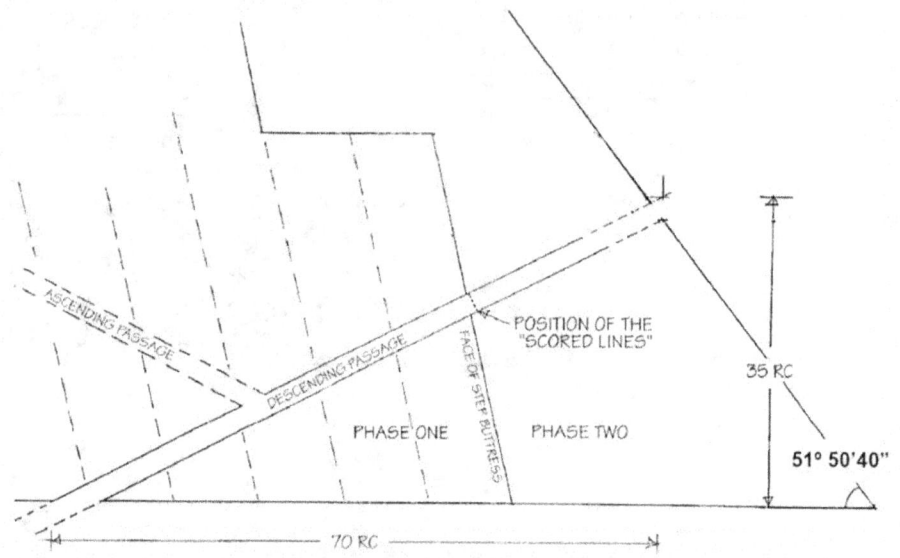

The Great Pyramid Partial Side Elevation Cross Section

The Queen's Chamber is also directly on the pyramid centreline and this position, could have been determined simply and accurately from the diagonals at the top of the core at 37RC.

The width of the Descending Passage is 2RC and perpendicular height 2.29RC (2RC+2palms) for both the below bedrock and masonry sections The vertical height appears arbitrary at 2.56RC, however when this dimension is added to the entrance height above the base of 32.44RC, it can be seen that the whole of the passage roof profile forms a triangle with a height of 35RC and a base of 70RC. These regular dimensions and shape could be used to create a template for the passage stonework, enabling prefabrication and also to determine the positions of both the passage entrance on the pyramid face, and the dimensions and location of the original rectangle (5.12RC X 2RC) marked on the base for the section of passage below ground.

From the rectangle marked on the pyramid base, masons would begin cutting and maintaining a 2:1 slope ratio for the passage. This could be achieved by first roughing out the face to an approximate size, followed by completion of the floor. Accuracy could be assisted by using a straightedge and a tool based on a 2:1 triangle and plumb line. The passage width could then be finished using a 2RC template and the ceiling finished last; its slope angle controlled using the same triangular tool, but inverted. The limited space meant that only one mason could be roughing out the tunnel face with others following in stages completing the tunnel to its finished dimensions. The direction of this underground passage is extremely accurate, differing by only 5mm. throughout its length.

Great care was necessary as the line of this passage affected all sections above, including the location of the pyramid entrance. A timber box section template, designed to fit exactly within the passage dimensions could have provided a final control of both straightness and size.

Assuming a completion rate of 20cms. per day, the whole of the angled and level section, together with part of the Subterranean Chamber could have been completed within one full year.

Once the construction of the reference core began the section of passage within the core also had to follow the line of the passage below ground. The passage floor blocks would be placed first on a base which maintained the 2:1 slope, followed by the blocks forming the walls and roof progressively sliding down each completed section.

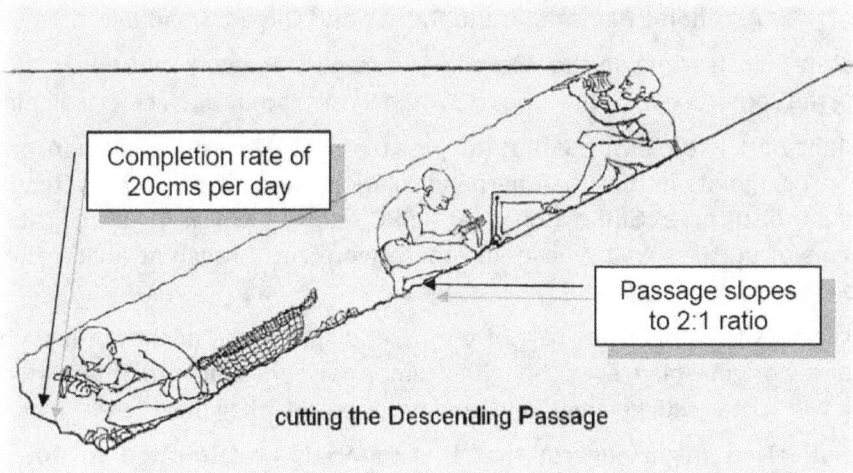

cutting the Descending Passage

Looking up the Descending Passage to the Pyramid entrance
Note the carefully cut stonework

Ascending Passage, Grand Gallery and Queens Chamber

There are odd features in the Ascending Passage stonework which was formed during the construction of the core to 37RC and might indicate a change in plan.

The stonework in the lower part of the passage is local limestone laid horizontally with vertical joints, rather than perpendicular to the passage floor as it is later. There are also three 'Girdle Stones' at 9.9RC intervals which might indicate the positions of buttress walls above and these were cut through at a later time to complete the lower section of the passage.

This suggests that the lower part of the Ascending Passage might not have been part of the original plan, even though its dimensions are similar to other passages, with a width of +-2RC, height 2.56RC and perpendicular height of 2.29RC.

Alternatively it might suggest that the materials prefabricated to form the passage were either not completed or delivered late and the construction of the core had to continue without them.

The 'unfinished' Ascending Passage in the Great Pyramid

The Queen's Chamber and Grand Gallery

The Queen's Chamber and related Grand Gallery and Ascending Passage is the most complex feature in the whole pyramid and comprises a chamber centred directly on the pyramid centreline, connected by a low passage to the Grand Gallery, which in turn is connected at its base to the upper end of the Ascending Passage.

The Grand Gallery formed from local limestone is close to 40 metres in length and rises at an angle of 26º12' creating a cross-section to a 2:1 ratio. This ratio would have provided an effective template when the stonework was prefabricated to specific dimensions outside the pyramid. It has a sloping floor with raised sections to each side and vertical walls. The corbelled ceiling is formed by arranging each course of blocks so they slightly overlapped the one below. The design of the Grand Gallery suggests that it might have been intended to be used as a holding chamber for blocks of granite, which would be released into the Ascending Passage after the burial ceremony in order to seal it. The priests carrying out this task would then have left the pyramid by the Well Shaft – a narrow and irregular passage connecting the Grand Gallery to the Descending Passage - before finally sealing the pyramid doorway in the façade. However, only three granite plug blocks are found at the bottom of the Ascending Passage.

The installation of the Grand Gallery was particularly challenging. The floor stones would have been the primary reference and had to be carefully placed with a 2:1 ratio slope. The remaining parts of the structure could then be slid downwards into place, automatically following this accurate foundation. The position of the first floor stone was therefore critical, as this would affect the position of all subsequent stones and the overall position of the Gallery within the pyramid.

At the lower end the floor contains a removable slab which disguises the entrance to the Queen's Chamber passage and must have been designed for this purpose. At the upper end the point at which the passage to the King's Chamber joins the Grand Gallery is rather clumsy. It employs a large block of stone known as the Great Step to make the change from a sloping floor to horizontal with the antechamber itself constructed from both limestone and granite with many blocks unfinished. It is noticeable that the front edge of the Great Step is aligned exactly with the centreline of the pyramid.

The Queen's Chamber is formed from local limestone and is rectangular with a pitched ceiling. Some parts of the stonework forming the walls and floor are unfinished.

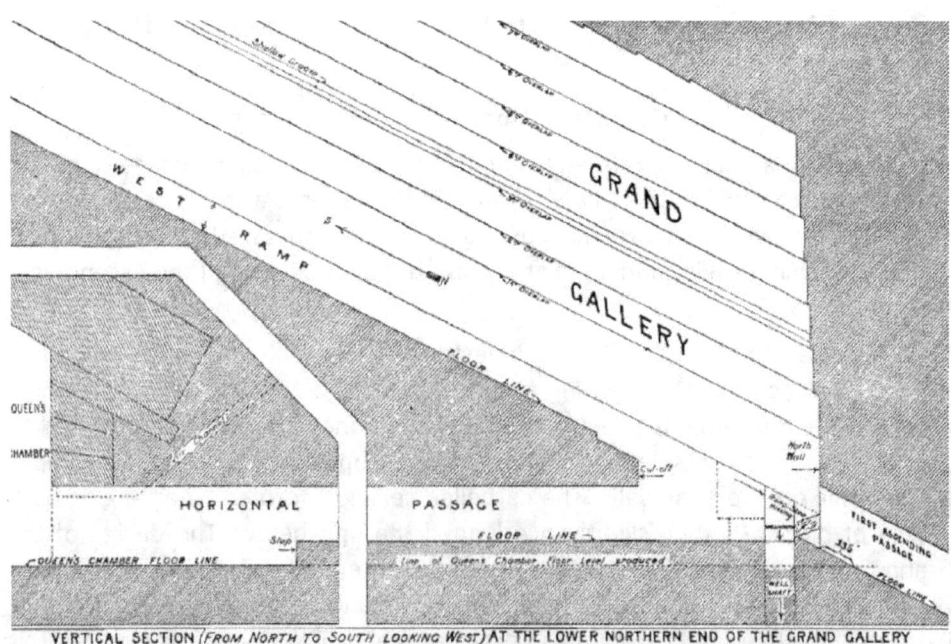

VERTICAL SECTION (FROM NORTH TO SOUTH LOOKING WEST) AT THE LOWER NORTHERN END OF THE GRAND GALLERY SHOWING THE HORIZONTAL PASSAGE, AND A SMALL SECTION OF THE QUEENS CHAMBER WITH ITS FLOOR-LEVEL PRODUCED TO ITS INTERSECTION WITH THE FLOOR OF THE FIRST ASCENDING PASSAGE.

VERTICAL SECTION OF QUEEN'S CHAMBER *(FROM NORTH TO SOUTH, LOOKING EAST.)*
SHOWING THE STEP IN THE HORIZONTAL PASSAGE LEADING THERETO:
THE NICHE IN THE EAST WALL; AND THE INCOMPLETE AIR CHANNELS.

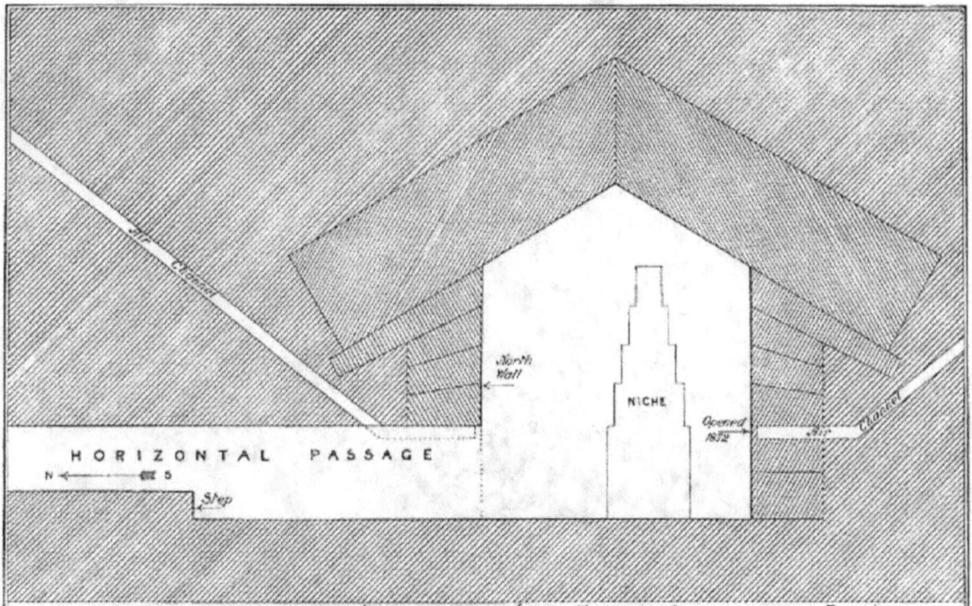

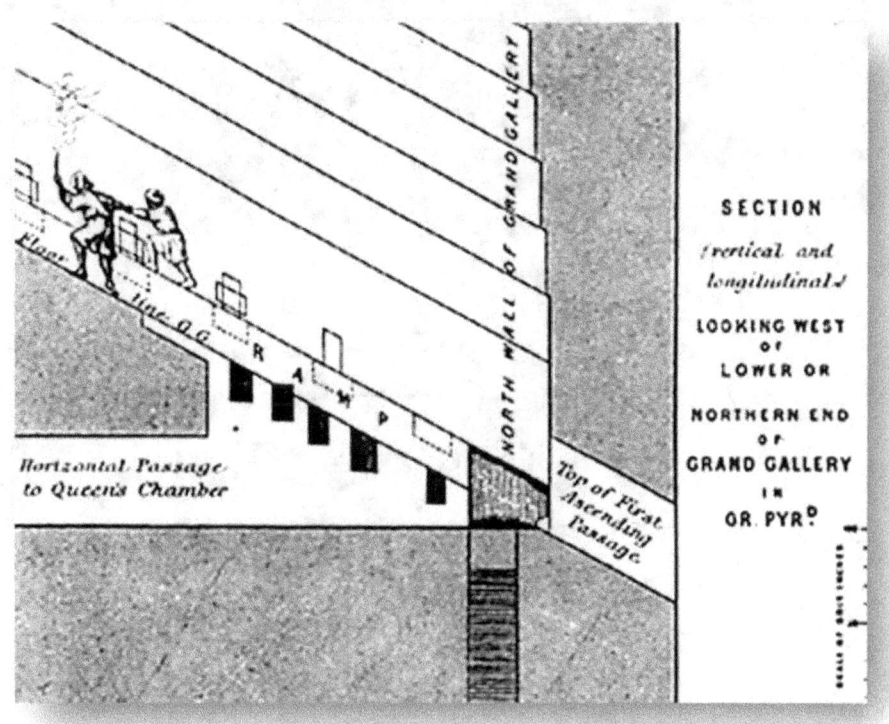

The Grand Gallery – looking up

....and down
The base of the Grand Gallery

The Antechamber and King's Chamber

The King's Chamber was placed at the upper end of the Grand Gallery with its floor at a reference height of 82RC (198RC from the apex). It is connected to the Grand Gallery by a small passage and Antechamber, the walls of which have grooves which might have held blocks of stone to act as a portcullis mechanism, designed to seal the burial chamber.

Antechamber South Wall

Antechamber North Wall

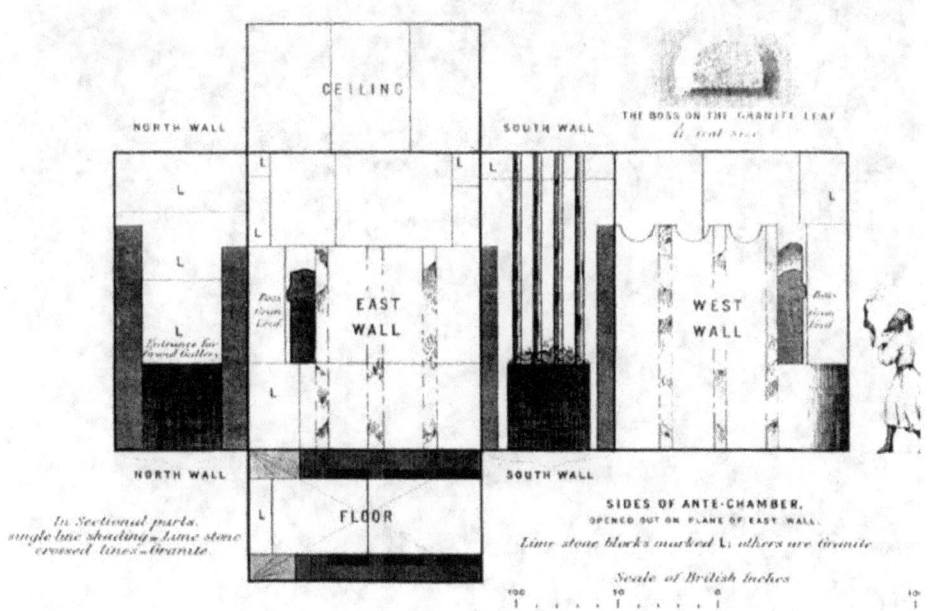

Antechamber details

(L = limestone)

The King's Chamber

The walls of the King's Chamber itself are constructed from 100 blocks of close fitting smooth granite, although the floor is incomplete. It has a length of 20RC, width of 10RC and significantly a height of 11.17RC formed from five equal courses of 2.29RC. (The base course is 0.29RC (2 palms) below the chamber floor). Also found within the King's Chamber is a granite sarcophagus on the pyramid centreline, but at right angles to the Great Step. Its height of 2RC is equal to the height of the entrance passages, which meant it had to be in place before the chamber was completed.

The King's Chamber – Great Pyramid

The Granite Sarcophagus – with Mr & Mrs Moreton Edgar

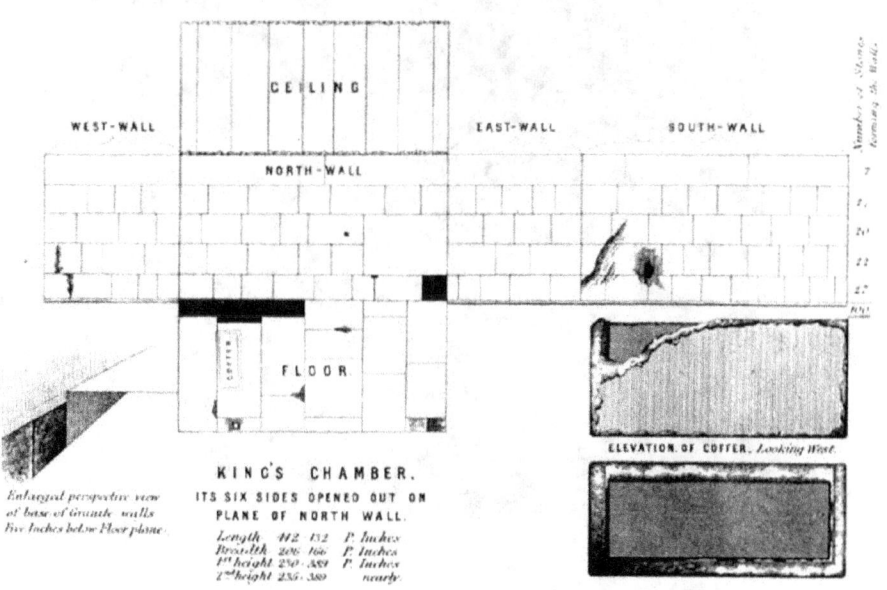

The King's Chamber Details

The Relieving Ceilings

The King's Chamber has an unusual feature known as the Relieving Ceilings These comprise 4 additional ceilings above the chamber roof itself, each separated by a small space, and are formed from large blocks of granite weighing up to 50 tonnes each. Other stone used to form the spaces between the ceilings is a mixture of limestone and granite. The Relieving Ceilings have this name because they are generally described as providing a means for reducing the weight of stone on the chamber roof. However it can be seen that they have no real structural purpose, as a vaulted limestone arch finally surmounts the whole system and they might have been placed as they are found, simply to consume these materials and to add over 15 metres to the height.

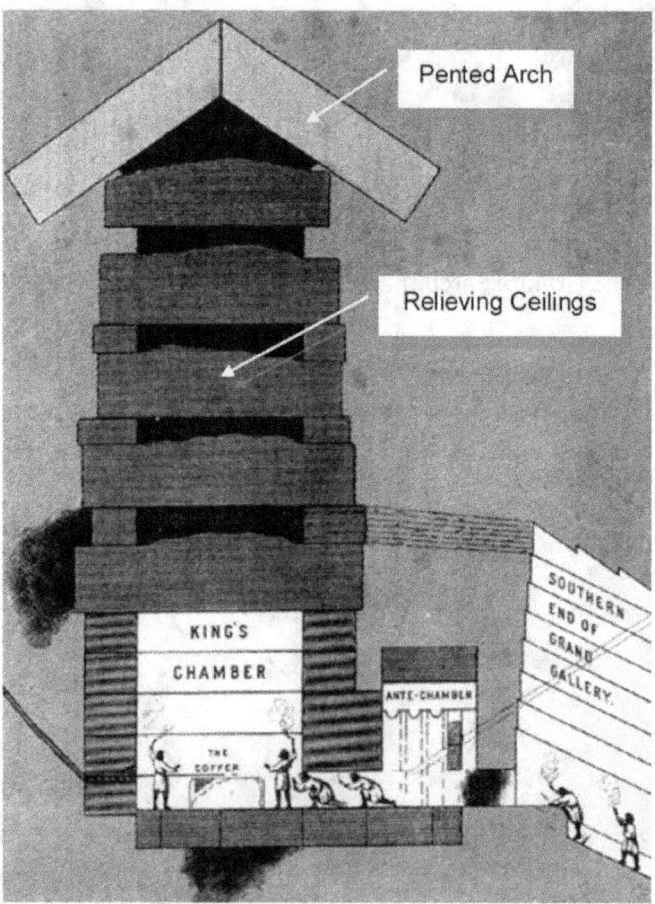

The King's Chamber – Cross section looking west

A Change in Plan?

A more logical answer might be that the long blocks forming the roof and the Relieving Ceilings above the King's Chamber were assembled from surplus passage roof and floor stones which had been intended to form the Ascending Passage in granite.

The dimensions and shape of these blocks is similar to those found forming the floor and roof of the Descending Passage and their total number would have successfully completed the roof and floor of the Ascending Passage including increasing its length, had they been used there.

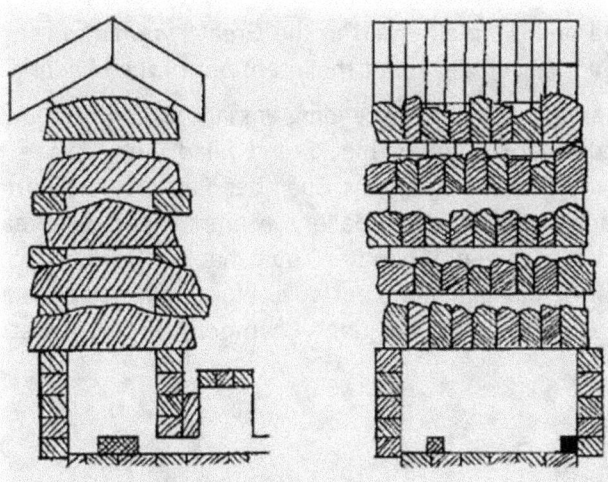

King Chamber and Relieving Ceilings

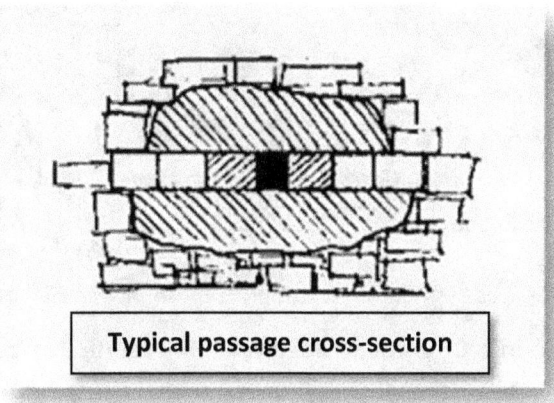

Typical passage cross-section

The height of the blocks used to construct the walls of the King's Chamber is also exactly the same as the perpendicular height of the Ascending Passage (2.29RC). Had these blocks been used here, their total would have provided enough stone necessary to complete the missing passage walls.

If it had been intended to securely seal the Ascending Passage with granite plug blocks throughout its length, it would have been consistent to form the walls, floor and roof from the same tough material.

If the design also included a granite portcullis mechanism for additional protection, then placing it at the entrance to the Queen's Chamber would have achieved this effectively, as all security sections below would be in granite and all ancillary sections above, in limestone.

Had this scheme been implemented in the Great Pyramid, no further comment would have been forthcoming about the intentions of the builders.

The walls, roof and floor of the Ascending Passage would be found formed from regularly shaped smooth granite blocks and filled with granite plug blocks to beyond the entrance to the Queens Chamber Passage, with an overall length matching the length of the Grand Gallery. A granite portcullis feature placed at the entrance to the Queens Chamber could have a 'granite leaf' blocking the raised section of the passage and 3 portcullis blocks lowered onto a floor covering the 1RC lower section beyond. The King's Chamber would not exist.

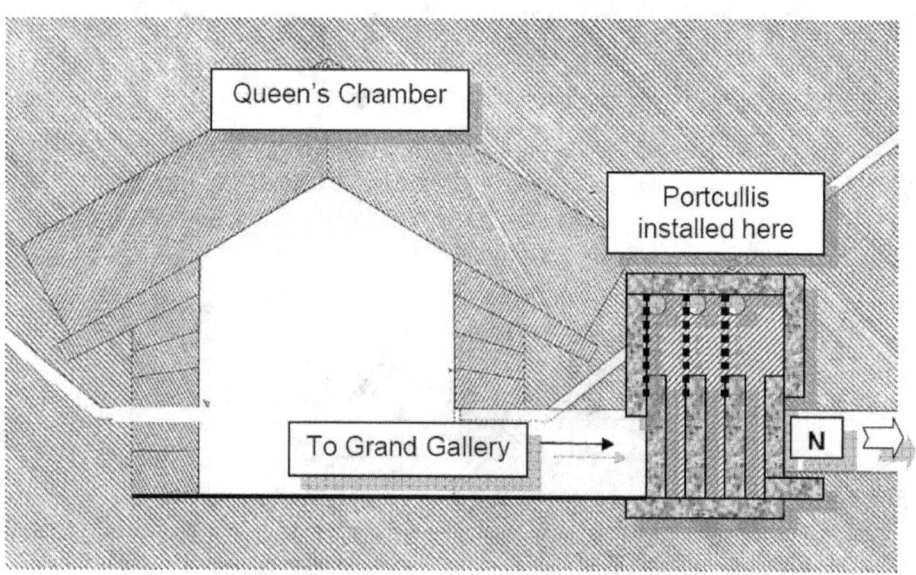

The Queen's Chamber Passage with Granite Portcullis

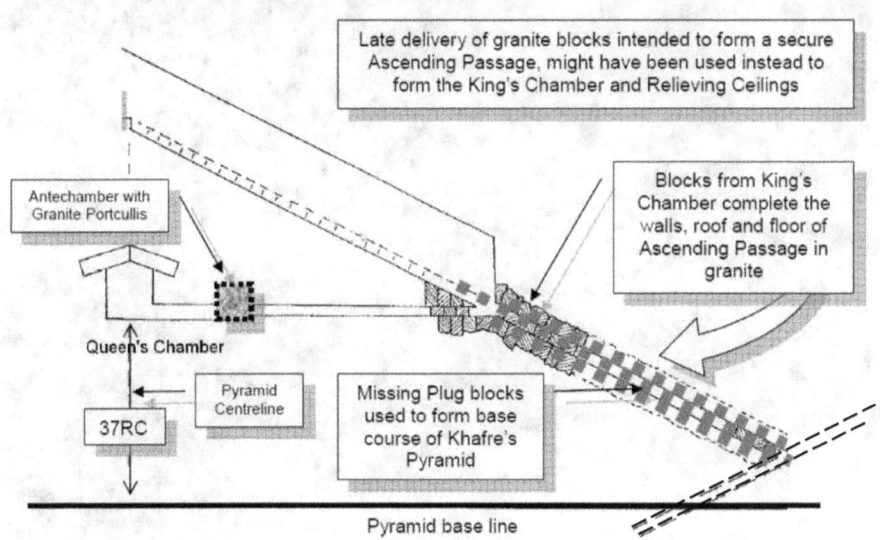

**The Queen's Chamber, Grand Gallery and Ascending Passage
in an alternative arrangement**

Compare this with what is found. An unfinished Ascending Passage formed in part by irregular limestone blocks, with one section cut later through the pyramid core. This crude passage is blocked at its lower end by granite plug blocks which represent only 20% of the capacity of the Grand Gallery holding chamber and were easily dug round to gain entry to the rest of the passage and all the pyramid chambers above. The Queen's Chamber and its passage at the base of the Gallery is unprotected, unfinished and without a sarcophagus. The Kings Chamber is found at the top of the Gallery, connected by a clumsy intersection through an unfinished Antechamber and portcullis, formed from blocks of both limestone and granite.

The King's Chamber floor is unfinished and the Relieving Ceilings above show evidence of contemporary cracking, settlement and mixed materials. No trace exists of the portcullis blocks themselves and perhaps they were never fitted at their upper position, being found instead either forming part of the Antechamber's or King's Chamber's floors.

It would be unreasonable and unfair to suggest that this was a planned design.

It would have taken a tremendous effort to both pre-fabricate and transport any granite blocks the 500 miles from Aswan to Giza. Those specified for the Great Pyramid were both large in number and dimensions, with many requiring accurate finishing. To complete an order of this size and on time would be a remarkable challenge and late delivery a distinct possibility, particularly so if the construction timetable of the pyramid relied on continuity. Any components which did arrive late and missed installation in that section of the pyramid for which they were intended, would therefore automatically become surplus. The options available to the builders would be to either discard them, or use them elsewhere. They would also need to formulate an immediate solution, which would replace the missing materials and maintain something of the original plan, even if this meant employing inappropriate materials and construction methods.

It might be that the King's Chamber and its Relieving Ceilings were created only as a consequence of a decision to consume surplus blocks, which had originally been intended to provide a competent security system for the pyramid interior, but arrived too late to be installed in their designed location. The design chosen was far less impressive than the corbel roofed chambers found in earlier pyramids which would have been possible with the blocks available, which might confirm this compromise.

The effort required to implement this change in plan would support a view that pyramid building was also an economic event, with all work/product carefully catalogued and rewarded. This condition would force the builders to ensure that all significant material was consumed, even if this action produced a less than perfect result. Some of the internal features in the Great Pyramid might therefore appear as found only as a consequence of unforeseen events, which had forced the builders to abandon their original design and replace it with a creative compromise.

This would explain the existence of the King's Chamber and Relieving Ceilings themselves and the inappropriate materials, poor quality of finish and function of the Ascending Passage.

Further evidence for passage linings in granite is clearly demonstrated in Khafre's Pyramid, with a granite entrance passage of perpendicular height 2.29RC, extending to just beyond its granite portcullis. Menkaure's pyramid also has a granite entrance passage and portcullis. As neither had plug blocks, the effort to form them in this durable material was wasted and might be evidence of other compromises. It appears that the bulk of granite block production at Aswan was to specific dimensions and shapes. Some matched typical passage floor and roof blocks being over 10RC in length and others were to two height specifications. One was consistently to a height of 2RC+2 palms (2.29RC), this being the perpendicular height of pyramid passages, whether formed in limestone or granite and blocks matching these shapes and dimensions are found in the Relieving Ceilings and the walls of the King's Chamber in the Great Pyramid.

Others were finished less carefully to heights and widths close to 2RC and these would therefore fit within pyramid passages and effectively plug them.

It appears that many of the missing 2RC blocks which could have been used to plug the granite passages in the other two pyramids at Giza and completely fill a granite lined Ascending Passage in the Great Pyramid, might now be found forming the base course of Khafre's pyramid and the first 16 casing courses of Menkaure's pyramid.

Granite casing Menkaure's Pyramid....unfinished.

Installation of Passages and Chambers

A difficult and time-consuming aspect of assembling each of the internal features would be the placing of the huge blocks involved. However as each chamber is placed at a reference height and construction work had moved from the core to the facade, the workforce at least had the time and space to deliver the materials and carry out the work. Animal power could also have been employed during these periods. As each section of the chambers and passages was completed, the surrounding core stonework would be raised to form platforms together with any necessary ramps, to ensure that the heaviest blocks could be dragged horizontally into position. The most difficult blocks to place would be those over the Queen's Chamber and the Relieving Ceilings. The Queen's Chamber has a cantilevered arch and the inner ends of these blocks would have dropped into position under their own weight when the temporary supports below were removed.

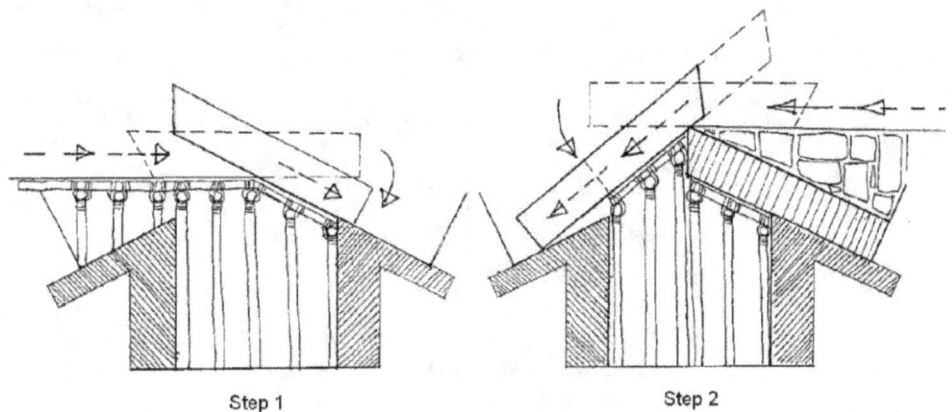

Step 1   Step 2

Installing the Queen's Chamber Roof Blocks

The King's Chamber ceiling and the Relieving Ceilings above would have been dragged horizontally into place from access platforms at the appropriate height. The vaulted arch above the Relieving Ceilings is not cantilevered, but could also have been installed using a similar method, by ensuring that all major movements were horizontal. Holes are found in the tops of the relieving ceiling blocks immediately below, which might have contained props used to form a scaffold, which would have supported the beams as they were dragged into position.

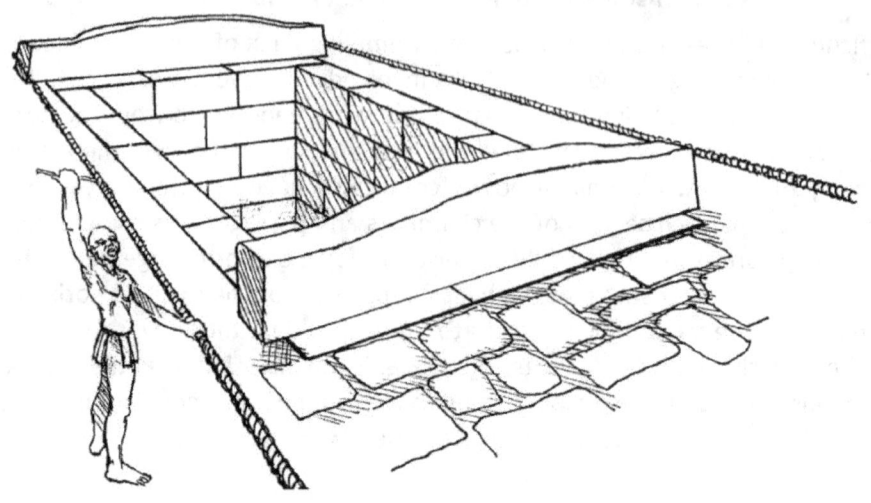

Fitting the King's Chamber ceiling

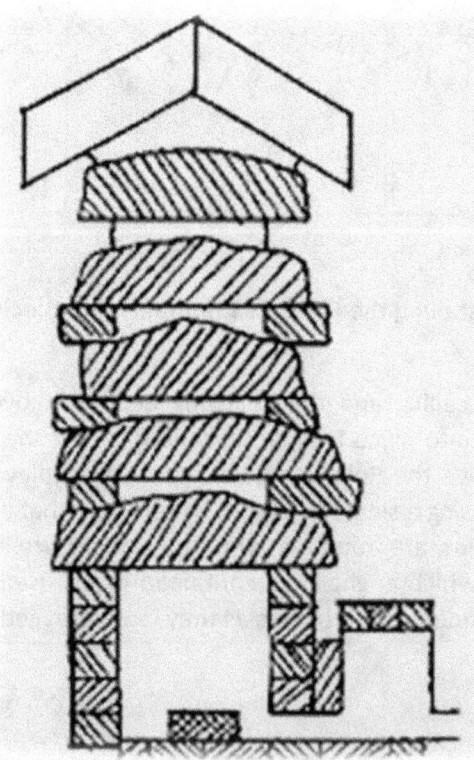

The Relieving Ceilings with Vaulted Arch above

The Air Shafts

A final set of features, which might also suggest the late and unplanned addition of the King's Chamber, are described as the Air Shafts. These narrow box section shafts rise at various angles from positions outside the walls of both the King's and Queen's Chamber. Their function has been described as a symbolic means for the Spirit of the Pharoah to have access to the exterior of the pyramid and would be 'opened' after internment of the body. If this is true, the Queen's Chamber must have been intended as a burial chamber for some time before a decision had been made to add the King's Chamber otherwise Air Shafts would not be found there even though they remained 'unopened' until recent times. It has been suggested that they might also have been aligned to stars, even though each has a horizontal section before turning and travelling up through the core. Legon has shown that the straight section of each shaft also takes the shortest route to the pyramid exterior and it is for this compelling reason at least, that for whatever purpose the Air Shafts might have served, star alignment does not appear to be one of them

.

An 'Air Shaft' entrance in the King's Chamber wall

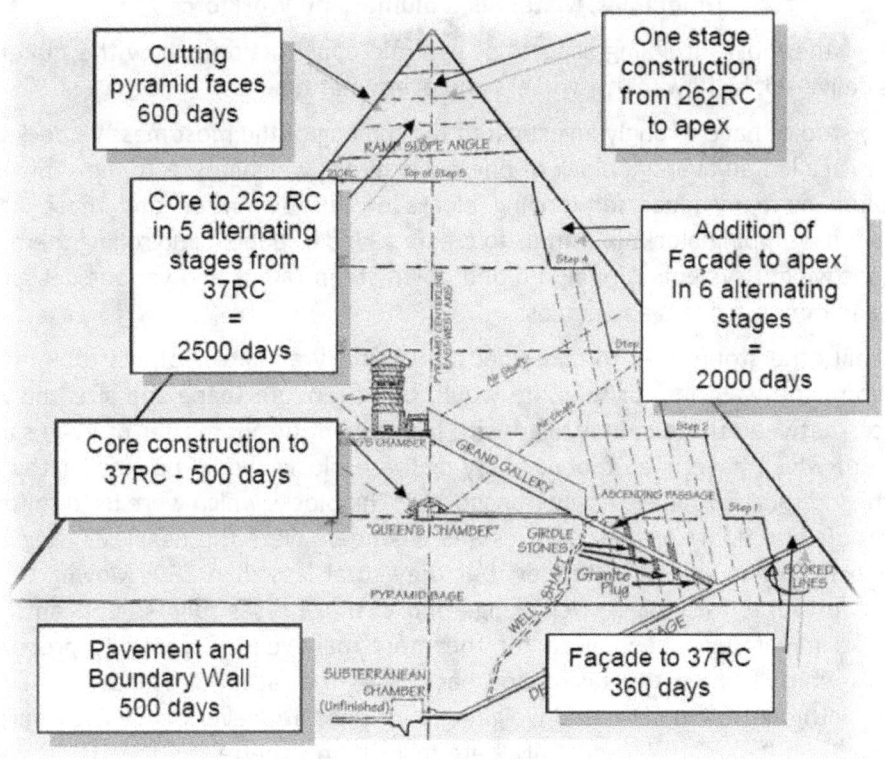

Great Pyramid Cross Section
Looking West

A Construction Timetable – 10000 workers

Core construction 5 stages	= **3000 days**
Façade Installation 6 stages	= **2360 days**
Cutting pyramid faces	= **600 days**
Pavement and Boundary wall	= **500 days**

Total = 6450 days (17+ years)

Satellite Pyramids and other ancillary works constructed during main construction phase

Timetables, Materials, Volumes and Workforce

One of the most intriguing aspects of pyramid construction is how the material was delivered, by how many workers and over what time.

Many studies have wrongly investigated the moving of the most massive bocks or what is called an average block of one cubic metre weighing 2.6 tonnes This has resulted in many ideas for moving blocks of this dimension and these have included wrapping blocks in frames to create a kind of bobbin and rolling them, or counterweight systems running up and down steep ramps and various design of tracked ramps, trolleys and cranes.

In reality the stone used to construct the Great Pyramid and others took many different forms. Stone for the core would be of random shape and size and the spaces between these stones filled with large quantities of mortar or even sand, both of which have been found. Many of these blocks would be smaller than 1 cu.m. and therefore more easily transported. The blocks which were used to form the roofs and floors of the passages and chambers were the heaviest, weighing between 20 and up to 50 tonnes, but they total less than 350. Moving these blocks presented the biggest challenge and in these cases rollers, lubricant and animals might have been used for the most massive. The carefully prepared blocks which formed the casing and backing of the façade are heaviest at the base, with one still in place and weighing 20 tonnes. However of the 210 courses in the Great Pyramid's façade, only 9 are taller than 1 metre and the majority are closer to half a metre.

The vast majority of blocks forming the Great Pyramid, including the façade weigh less than 1 tonne, with huge volumes of infill formed from material of random size and shapes.

The total weight of stone used to build the Great Pyramid is commonly calculated as the volume of the pyramid multiplied by 2.6, the density of solid limestone, giving a total of 6,740,000 tonnes. The reference core which comprises more than 85% of the volume is not solid limestone and its more random composition might have a density closer to 1.5 and a total weight of 3,300,000.

In this case the weight of the Great Pyramid falls by 2,400,000 tonnes or 35% to 4,300,000 tonnes.

Any timetable for the construction of the Great Pyramid must take account of how much weight an individual worker can be expected to move routinely, in what form and to where. Teams of up to ten men would have been enough to move the majority of blocks, with smaller groups engaged in taking infill material both for the pyramid and its associated ramps and platforms. The design presented here ensures all blocks are positioned without lifting. The largest blocks, much lower in number, would have involved much larger teams of workers, perhaps in rotation to maintain momentum.

Carrying material is an efficient method of moving lighter material and much of that for the core, infill and ramp construction, might have been in weights manageable by one or two men and as low as 20kg each. Studies, which have investigated heavy block-moving methods, have found that up to 100 kgs can be moved by one man up a gradient of up to 5%, but an allowance must be made if this is to be achieved consistently over distance.

The time available to carry out the delivery of some of the most difficult and heaviest blocks to the top of the core at 37RC (20m) and others to 82RC (43m), was however throughout the whole period of façade installation to these heights and with access from a broad perpendicular ramp and animal assistance. This would include a total of 150 massive blocks used to form the Relieving Ceilings above the King's Chamber.

A further 200 limestone blocks of similar shape and size were used to construct the Descending Passage at a much lower height and had to keep pace with construction of the reference core taking place simultaneously around it. Once the passage floor was installed accurately, the sides and roof could be placed relatively easily by sliding them into place using the floor as a runway. Access here would be directly from the side of the perpendicular ramp alongside.

The volume of material to build a solid pyramid reduces significantly as the pyramid grows in height, but the effort and time required delivering material increases. For example the first 20 metres of the pyramid height consumes over 30% of the total material and the final 30 metres only 1%. Also a given number of workers might be twice as effective at the base as at mid-height and twice as effective in constructing the core as fitting the façade.

However, the construction methods described here which exploit the virtual apex method provide features, which positively affect the overall timetable as many of these processes are taking place simultaneously. As material is being added to the reference core, prefabrication of the façade is taking place. As the façade is being added to a reference height, the internal features are being assembled on the reference core. At all heights a platform level with the top of a complete course is available to take blocks to any part of the pyramid.

It is not surprising that those studies which, having assumed pyramids were constructed in a single continuous process from 1cu.m. blocks of solid limestone and laid without accurate references or comprehensive access, continue to produce implausible results.

Constructing the Core to 37RC

The lower part of the reference core has a working area of close to 50000 square metres, a low height and good access from a broad perpendicular ramp with a slope of less than 5%. In one scheme every one of 10000 workers would take part in delivering an average of 20kgs of material each hour to a total of 2000 tonnes per day. The on-site movement of men and materials would be critical for this schedule to be successful, as a delivery cycle of one hour would require the whole workforce to enter and leave the work site 10 times each day. This would tend to support a delivery system based on flexibility and ease of movement rather than one using a lesser workforce struggling with massive loads. In this model for example, up to 50 men would be available to move a block weighing even 1 tonne. They could be employed as 5 teams of 10, moving the same block in rotation, or any similar combination with an aim of achieving maximum efficiency for their individual daily targets. Any surplus labour would also be available to routinely carry baskets of smaller infill material. The installation of the Descending Passage would take place simultaneously with construction of the core, with the overall schedule unaffected as all the blocks for this feature would have been prefabricated. On the basis of this scheme the completion of the core to 37RC would have taken 500 days with each member of the 10000 workforce delivering a total of 135 tonnes of material over the whole period. Compare this conservative estimate with that of the more modern 'Navvies' expected work rate of 20 tonnes per day. The time available would allow the Descending Passage to be installed at a rate of 1 linear metre every 3 days.

When the core had reached a level platform 37RC (19.4m.) above the base, 1,000,000 tonnes of material had been delivered, representing more than 20% of the total weight for the whole pyramid. The majority of this material would be in the form of rough-hewn blocks, with substantial quantities of infill mortar and even sand. The exception was the prefabricated blocks used to form the floor, walls and roof of the Descending Passage. An additional 45,000 tonnes of material would have been required to build a perpendicular ramp with a 5% gradient, 5 metres wide to 37RC, during the same period.

Adding the Façade to 37RC

The material used to construct the façade was closely fitting limestone with a density of 2.6. Courses close to the base were the tallest, some having blocks weighing up to 20 tonnes. The workforce would have to deliver and place 360,000 tonnes of stone to complete the facade to the height of the core at 37RC. A further total of 50,000 tonnes of material was required to maintain the perpendicular ramp and the external platforms at each course height.

10000 workers would have to deliver 1000 tonnes of material every day - the majority prefabricated - with each taking part in moving an average of 20kgs. to the pyramid in every two hour period. This represents 20 typical façade blocks with an average weight of 5 tonnes at these heights, arriving at the pyramid every hour. Each would be moved by teams of up to 250 available labourers divided between the four faces.

At the lowest levels feeder ramps would provide multiple access points to the platforms surrounding the casing, which was raised in height as each course was completed. When the maintenance of the feeder ramps became inefficient at higher levels, the original perpendicular ramp would be brought back into use by connecting its upper end to the platforms at each course height.

From this point on, blocks arriving at the top of the ramp would be taken sequentially around the platform, to their position on one of the four faces. Casing blocks would have their tops levelled and the platform raised to this height as soon as they were laid, leaving only the marking of the face edge on them when the whole course was in place. In this scheme the installation of the prefabricated façade to 37RC could have been completed in 360 days.

A telling statistic can be taken from the total length of casing of the Great Pyramid which had to be laid accurately as thousands of tonnes of other stone was being placed elsewhere. The façade contains a total of 96000 linear metres of casing, 18000m forming the first 21 courses to a height of 37RC.

For the façade to be completed to this height in 360 days, 50m of white casing had to be accurately laid each day. If external access is available to all four faces, as in the present case, this average is less than 13 linear metres to each face, each day. The installation process would be significantly helped with prefabricated blocks, access from a level platform and an accurate and rapid means for marking the face edge onto a completed course.

If this rate were maintained for the façade courses above, which were generally formed from blocks of lower height and a weight of less than 1 tonne, a total of

2000 days would be required to complete the whole of the façade in its alternating stages to the apex.

The total of 860 days to move and fit one third of the whole pyramid volume is based on a reasonable number of workers and the effort they have to provide in moving typical material and working to a design based on the Virtual Apex method and prefabricated material in critical areas. A further 3000 days would have been necessary to complete the intermediate reference core stages giving a timetable of 5360 days of alternating construction stages to complete the whole pyramid to the apex. A further 600 days must be added for cutting the four faces of the pyramid from the apex to the base, including ramp removal and an extra 500 days to complete the pavement and boundary wall closest to the pyramid base.

With partially prefabricated blocks laid to a reliable guide for the façade and internal features, random stone used for the core and external access to all faces of the pyramid from a ramp of low gradient throughout, at least a routine method for the delivery and accurate installation of materials is provided.

The complete ramp and platform system, including the perpendicular ramp to 37RC, required almost 250,000 tonnes of material to take it to the apex, an average of 100 tonnes per day. Some 1000 workers would have to be engaged on the task of maintaining the ramp and platforms and without interrupting the flow of other men and materials. The ramp/platform design presented shows how the bulk of this material is used to raise the platform to the top of a laid course, which begins at the point on the pyramid furthest from the top of the ramp. This work takes place immediately after a casing block is placed and does not interfere with the access ramp itself or the placing of other blocks. The total material to take the end of the spiral ramp to each new platform height is less than 20 tonnes.

Quarrymen, Masons and Support.

Quarrymen at Giza could have been working during almost the whole construction period, to free an average of some 1150 tonnes (500cu.m) of material each day. 2500 quarrymen and labourers working continuously would be required to maintain this target, each freeing less than 1cu.m, every 5 days.

Of this material 60 cu.m would have been prepared to act as backing stones between the core and the casing.

An average of 35 tonnes (14 cu.m) of white limestone each day was required to keep pace with the prefabrication and installation timetable of the casing blocks for the façade. (96000m X 0.7m X 1m X 2.6 ÷ 5000 days)

If a mason took 10 days to free 1cu.m, then 150 could meet the daily target.

A further 650 masons would be at the Giza site, 150 prefabricating the casing blocks and 500 the backing blocks at the same pace as the stone arrived from the quarries.

During the period the site was being prepared and the core built to its first reference height of 37RC, enough stone could have been quarried, delivered and prefabricated by these teams, for the whole of the facade to the same height.

The workforce at Aswan, preparing the granite blocks used in the interior of the Great Pyramid might have totalled 500, as 1600 days (including site preparation) was available for them to prepare and transport the material to Giza, before the core had reached its first reference height of 37RC, where this material was to be deposited. If this material had been intended instead for constructing the Ascending Passage and its related plug blocks, as suggested earlier, the first of these blocks would have to have been ready for installation up to 500 days earlier and if late might force a change in plan!

Support workers would also be required throughout the project and would include, mortar mixers, water carriers, tool sharpeners, copper miners and smelters, drawn from a general labour pool.

The total workforce in this model is therefore close to 15,000, completing the Great Pyramid in 6450 days.

Adjustments can be made to any of these elements, which might result from any block moving experiments which attempt to duplicate these conditions, but any proposed alternative method of pyramid construction, material installation, or workforce numbers, must be tested against the casing and ramp/platform statistics and timetable presented here.

Conclusions

This practical solution is based on the special social and geographic conditions which might have provided the geometry, alignment skills, labour force and materials, enabling the Ancient Egyptians to cut, shape and place stone accurately to form solid pyramids of different shapes.

The design and method of construction described does not include any machinery or tools which did not exist at the time and leads to pyramids of the dimensions and shapes found. The evolution of the design process might be identified from the shapes and style of individual structures.

The Stepped pyramid was enlarged around an existing stepped building, itself a development from earlier burial chambers and might demonstrate the discovery and application of geometry as a means to accurately control the shape of large structures. The Meidum pyramid maintained the stepped style in its first stage, but appears to have been revisited to add a smooth casing after the Bent and Red pyramids had been completed with this feature.

The Bent Pyramid was the first attempt at building a pyramid with a smooth casing and the shape chosen appears to exploit the virtual apex method in its simplest form, with a height:centrepoint-corner ratio of 1:1. The change in shape for its upper section employs a different virtual apex ratio of 2:3 for the first time which might have been a response to structural problems although the diagonal cross-section dimensions provide an elegant design for a dual shape pyramid 200 RC high. The Red pyramid which followed employed the upper section ratio successfully throughout its height. The smooth façade added to the revisited pyramid at Meidum was to a height:centrepoint-corner ratio of 9:10, which was to be repeated in the next structure – the Great Pyramid itself.

Each of the earlier structures has elements, which might be seen as less than ideal, with each new project demonstrating attempts to rectify those identified.

A smooth facade preferred to a stepped profile, a low angle of slope more stable than steep; a single steeper slope more pleasing to the eye.

The Great Pyramid is a structure, which marks man's first major technological success. It demonstrates how new materials and techniques will be used in any age to overcome what is seen by some as an impossible task and to create something which encapsulates this endeavour and success.

The addition of extra steps and a smooth façade to the Meidum pyramid might suggest that knowledge of the relationship between 70 and 99 as a substitute for $\sqrt{2}$ was known and exploited here, as the centrepoint was covered throughout construction. If the side length of the extra steps were measured at reference heights using the 70:99 relationships, it would have provided a means to determine the virtual centrepoints and from these the corner edges at selective heights.

The division of the Royal Cubit into seven parts might further confirm knowledge of the relationship between 70 and 99 and indicate that reference squares could have been employed for marking pyramid bases by providing a known dimension from each corner to the virtual centrepoint.

The layout of the Giza site appears to have employed a single diagonal reference line crossing the most level section of the site with a series of points along its length defining the position of two sides of each pyramid.

The design of integrated platform and ramp described meets the essential requirements of providing a low slope, requiring a minimum of material to maintain it. The use of a ramp of this design is only possible when the alignment of a pyramid applies the virtual apex method which uses the corner edges as the primary reference and does not require completed lower sections to remain visible.

Compelling evidence for the use of the diagonal : height ratios as a construction device can be tested by attempting to build even a small pyramid accurately using any other method, when the apex is not in place as a reference.

It has been suggested that large pyramid building reflected the economic prosperity in Egypt at that time, which provided the time and labour force for this endeavour. However, any enterprise of this magnitude could become the momentum of an economy, ultimately forcing the activity to be maintained for this reason alone. Pyramid building was a labour intensive, low-tech enterprise employing large numbers of workers in routine tasks, yet many of the discoveries the Ancient Egyptians made during the Pyramid Age have had to be re-discovered in modern times.

They were using the first kind of production line, pre-fabrication, shift working and just-in-time parts availability. Their construction techniques include adding a facade of a different material from the internal structure and even topping out ceremonies. They could also have discovered an alternative way of forming a right angle and partial diagonal in one simple construction to form squares with a known virtual centrepoint.

This discovery might have been made only because the river Nile flooded each year, which required regular field marking geometry to be carried out and on a large enough scale to make the observation.

Their decision to build large stone structures which applied these discoveries might have been for the same reason, as a large workforce would be idle during the same period of inundation.

Of the possible solid three-dimensional shapes available, a pyramid is the simplest. For a large pyramid however it only becomes possible to build it, if it is understood how the shape can be accurately formed without the apex for reference. Equally vital is to know how to provide comprehensive access for the delivery of material to all parts of the structure, including the exterior.

The evidence in their pyramids illustrates that the Ancient Egyptians had this knowledge, with its foundation in simple geometry and its effective application. It was in the Great Pyramid the Ancient Egyptians confidently brought together all their expertise in building large stone structures, by planning the tallest and most carefully made pyramid of all.

Clarke and Englebach suggested that the standard of Egyptian masonry declined after the Great Pyramid. This appears true when studying the finish of the backing blocks in the other pyramids at Giza which are of a much poorer fit. Also pyramids constructed later still, some of which had a mud brick core, are now almost complete ruins. However, what is also true is that the method of construction became more economical with labour and materials, but still achieved the objective of a smooth sided accurately shaped pyramid.

After the success of the Great Pyramid it would have been noted that the backing blocks did not need to be so close fitting in order to support the casing. Also as the core was only to provide the references for the pyramid corner edges and was protected by the façade, it could be built from a less durable and more manageable material such as mud brick.

In a sense this is an improvement in the construction method, providing the outer casing was maintained to the same high standard. If the casing stonework had not been removed from any pyramid, it is reasonable to assume that they would all be in almost perfect condition today.

Paradoxically there might also be a constant debate discussing whether or not to break into and explore these Ancient Monuments.

A question which might be asked is what reasons did the Ancient Egyptians have when choosing the dimensions for their pyramids. If there were a decision to include a seked in the profile of the largest pyramids, only two of those at Giza appear to demonstrate this with any clarity.

It might be the case that later generations which first reported this feature and unaware of the building method, were simply seeking to identify some kind of relationship between each structure, much as others do today.

Perhaps the Ancient Egyptians, keen to keep their methods secret, created the seked deliberately as a red herring, to divert the attention of others from looking too closely at the diagonal cross sections.

It appears to have worked! Even today many Pyramidologists continue to concentrate their efforts on fruitless attempts to discover ever more exotic meaning in these structures.

How it would amuse the Ancient Egyptians if they knew of these efforts which, having most probably exceeded their own, are no closer to discovering the simple truth.

The reality is, despite Pyramid dimensions being different in every case, each conforms to the demands of applying the Virtual Apex and Virtual Centrepoint geometry to build them.

The versatility of the Virtual Apex Method is such, it can be applied to form any shape of 'pyramid', with a base of any number of sides, of any length including unequal, and with each corner edge running straight to an apex at any chosen height and directly above any fixed point on the base.

The Ancient Egyptians chose to make their pyramids with shapes and dimensions which exploited the method in its simplest form.

Supplements

Finding π in the dimensions of the Great Pyramid

It has been noted that the Great Pyramid height and side dimensions form a relationship close to Π as 440 ÷ 280 = 1.5714.. X 2 = 3.1428.. Petrie even suggested that the side dimensions might have been intentionally 439.82RC (diagonal 622 RC) which is the average of his survey, as this gives an even closer fit.

However, the dimensions chosen for this pyramid, might only have been a result of the Royal Cubit having 28 fractions and the need to apply a simple height: : centrepoint-corner ratio in order to construct it.

Any pyramid which has a centrepoint-corner dimension one ninth greater than the height would have exactly the same proportions, and shape as the Great Pyramid, and therefore the same relationship with Π in its height and base side dimensions. The Meidum pyramid is an example.

The way this relationship appears almost exactly in this pyramid can be explained because its height is a multiple of 14 and the centrepoint-corner dimension is an additional one 9th of this. If these numbers are expressed in a form which reflects how they are used to provide the pyramid shape, they demonstrate how and why this relationship exists at the base.

Great Pyramid Dimensions:

To simplify, the height and side lengths are divided by 20

$$\text{Side} = 440 \div 20 = 22, \text{ and Height} = 280 \div 20 = 14$$

The side dimensions and diagonal of a square are related by √2 therefore:

$$22 \text{ can be expressed as } 14 \times \frac{20}{18} \times \sqrt{2} \text{ (or 99/70)}$$

$$\text{and 14 can be expressed as } 11 \times \frac{18}{20} \times \sqrt{2}$$

By eliminating common features:

$$14 \times \frac{10}{9} \div 11 \times \frac{9}{10} = \frac{140}{9} \div \frac{99}{10}$$

$$= \frac{1400}{891} = 1.571268\ldots \times 2 = \mathbf{3.143..}$$

A simpler way of calculating 'Eπ' based on the above is:

$$\sqrt{2} \times \frac{20}{9}$$

In Ancient Egypt a method for calculating Pi might therefore have been:

$$\frac{99}{70} \times \frac{20}{9} = \frac{22}{7} = 3.142$$

Much later texts describe a different and less precise method:

$$3 + \frac{1}{9} + \frac{1}{27} + \frac{1}{81} = 3.16$$

A Great Pyramid formula can also be presented:

$$\text{Height} = \sqrt{\left(\frac{\text{base side}^2}{2}\right)} \times \frac{9}{10}$$

Or: $$\text{Base side} = \sqrt{\left(\frac{\text{height}^2}{2}\right)} \times \frac{20}{9}$$

It is also possible to describe a solely geometric construction process for pyramids with shapes which have whole number height:centrepoint-corner ratios.

Let x = the given height = AB

1. Mark a line AC on a level base with AC = 2AB and midpoint B.
2. Construct a right angle perpendicular to AC at B.
3. Mark 2 points D and E on the line at right angles to AC at B, with dimensions = AB.
4. Form a square AECD with diagonals AC and DE. This square is the Reference Square.
5. Divide a line of length = AB geometrically into equal fractions based on the height:centrepoint-corner ratio. (e.g. Great Pyramid = 9)
6. Add one fraction, to extended diagonals AC and DE to points Aa, Ee, Cc and Dd.
7. Form a square AaEeCcDd. This square is the base of the Pyramid.
8. Begin construction of the reference structure on the Reference Square, AECD
9. Maintain a corner slope ratio of 1:1 (45 degrees).
10. Stop construction at any height y, which is less than AB. (say one fifth of AB)
11. Make the corners of the level reference square at height y, F,K,H,J. with diagonals FH and JK, and midpoint G. (Corners FKHJ are above AECD respectively.)
12. Divide a line of length = FG geometrically into equal fractions based on the same height:centrepoint-corner ratio.
13. Add one fraction, to extended diagonals FH and JK to points Ff, Kk, Hh, and Jj.
14. Take straight lines from Ff to Aa, Kk to Ee, Hh to Cc and Jj to Dd. These define the pyramid corner edges to height y.
15. Begin adding a facade from the pyramid base square Aa,Ee,Cc,Dd., to fit the defined corner edges.
16. Ensure the pyramid faces align to a straightedge or line connecting the corner edges at relevant heights.
17. Continue construction in this way to height y.
18. Re-start construction of the reference structure from square FKHJ at height y, to any new height z, (which is less than AB minus y), following the same procedures and order described in 9-18.
19. Repeat procedures 9 – 18 to an appropriate height below the apex P.

20. Place a vertical marker at the centrepoint of the final reference square, with a height equal to the dimension from a corner to its centrepoint. The top of the marker denotes the apex P.
21. Complete the pyramid to the apex P, aligning the corner edges with straightedges to apex P and faces aligned to straightedges connecting the corner edges.

Notes

1. At the base and any height above the base, the centrepoint-corner dimension of the reference square, equals the height remaining to the apex as ABP is a right-angled isosceles triangle.
2. Extensions to A-F, Aa –Ff, K-E, Ee – Kk, C-H, Cc-Hh, D-J, Dd-Jj all meet at apex P.
3. In the Great Pyramid, If $x = 1$ then: $x + (x/9)$ $X\sqrt{2} = 1.5713 = 3.1427$.

Great Pyramid - Course Heights of the Façade

The courses of the façade in this pyramid are of various heights with the tallest at the base. A pattern can be detected however, which has courses above the base diminishing in height until a new taller course restarts the pattern. These patterns continue to one third of the height, with courses above generally of similar height.

The reasons this pattern appears might be a consequence of the prefabrication of façade blocks described earlier. As the first blocks were brought to the site for preparation, they would be selected according to their height and arranged into groups, with each set having a total length necessary to complete one course. It would be inevitable that some blocks would become surplus as their total would not be enough to complete a course.

This would be particularly so at the base, where course lengths are longest. However as the course lengths diminished as the height increased, the total surplus blocks of a particular height would eventually be of sufficient length to complete a shorter course at a higher level, but also with a surplus. These taller blocks when in place would restart and repeat the course height pattern. Evidence for this might be found in the position of the third tallest course in the Great Pyramid which is at a height of 55RC.

At this height the total course length would have been the first which was short enough to allow all the surplus of the tallest blocks originally prepared for the base courses to be finally installed on the pyramid.

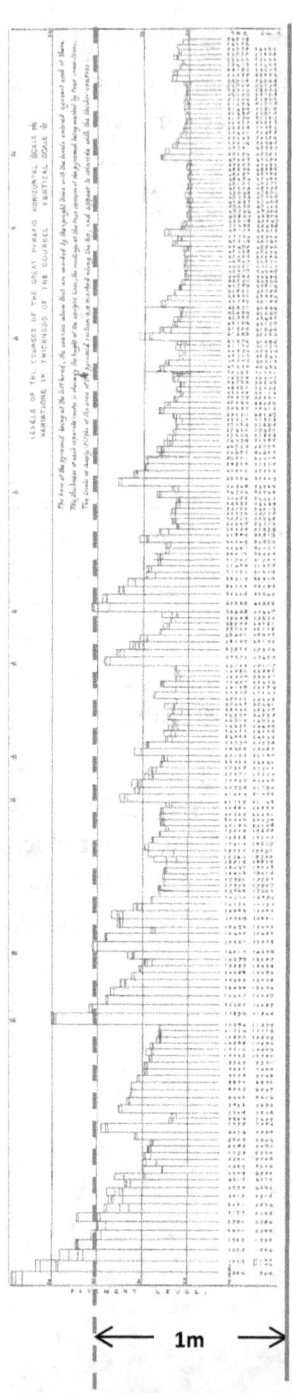

Façade course heights of the Great Pyramid
(Only 9 courses over 1 metre in height. Average height is 70cms)

The Advantages and Evidence of Two Stage Construction

There are considerable advantages in using a two-stage method of construction to build a pyramid as it divides the work into distinct complimentary sections and certain internal features confirm its application. The position of the Queen's Chamber floor is 41RC above the base, with the underside of these blocks placed on the top of Step 1 of the reference core at a reference height of 37RC. The King's Chamber floor is also close to the top of Step 2 which implies that the blocks forming this floor were also laid directly on the core at a reference height of 82RC. As the construction at each of these core reference heights moved to fitting the façade to the same height, ample time and space would be provided for the assembly of these complex internal features. Similarly, during construction of the core structure to each reference height time was available to prefabricate the blocks for the facade which would be installed later to the same heights.

The effect on a construction timetable can be significant when more than one activity is taking place in a given time period. This might double the production rate, or halve the workforce or be combination of both.

Other evidence of two-stage construction can be found from an analysis of the only markings visible within the whole pyramid. These are known as the "Scored Lines" and appear on the walls of the Descending Passage at a position 23.38RC from the pyramid entrance at floor level. They are marked on what were the first blocks of the passage, placed during the fitting of the facade, as they are 1RC outside the line of the outer buttress wall of the reference core, at a position 21RC above the base.

A perpendicular line taken through the passage floor at the entrance, intersects the line of the passage roof exactly 35RC above the pyramid base.

The whole of the stonework for this passage and the pyramid door mechanism could therefore be prefabricated on the ground to a triangular template of base 70RC and height 35RC, providing a slope with a horizontal to vertical ratio of 2:1.

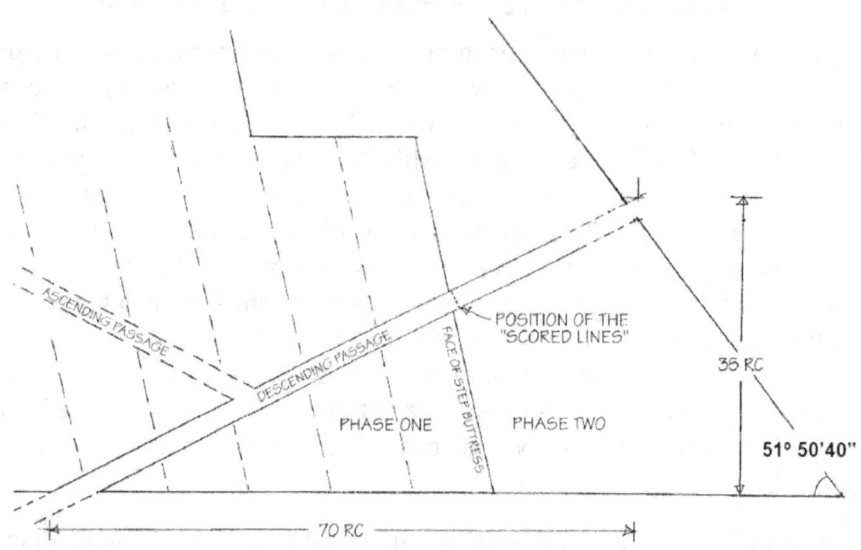

Great Pyramid - Side Elevation Cross section

The blocks, which formed the lower part of the passage, were installed during the construction of the core to the first reference height. Some mark had to be placed on the blocks left behind as a dimensional reference to ensure they would accurately connect between the passage blocks laid earlier in the core and the complex stonework of the pyramid entrance within the façade.

The "Scored Lines" could have provided this reference and be visible today, because the blocks on which they were marked, might have been placed in the wrong order. As the wall blocks were installed, by sliding them down the smooth passage floor, if those for the left side were placed on the right, and those for the right placed on the left, the lines would be visible on the inside, rather than covered on the outside.

Similar lines might also have been marked on the blocks forming the passage floor and roof, but as they could only be placed correctly these lines would be covered and therefore remain undetected.

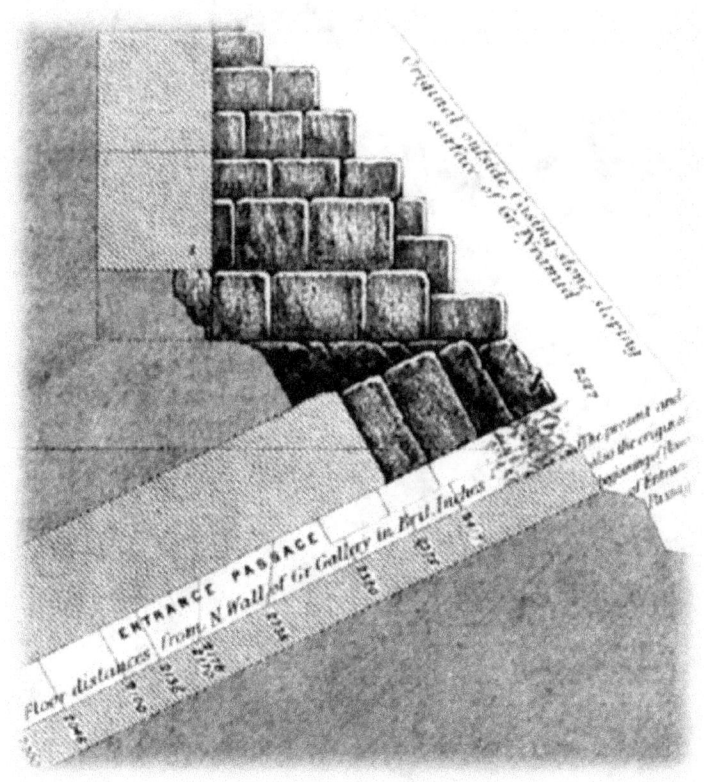

Great Pyramid Entrance

Adjustments to the Eiffel Tower

The builders of the Eiffel Tower were aware of the problem of directing an inclined corner accurately when there is no direct physical reference. Despite carefully prefabricated components using comprehensive dimensions taken from over 5000 scale drawings, they found it necessary to incorporate an adjustment mechanism into each of the four legs as they rose independently from the base. Adjustments were made as the components were brought together for the first time at a height of 57m, using hydraulic jacks at the base to provide angular movement and sand-boxes on a support scaffold, for vertical and horizontal movements. This problem was a result of taking a traditional approach to the design using the profile of a structure as the basis for its dimensions in both elevation and plan. Each leg is inclined to a 2 to 1 ratio in profile, but this simple geometry could only be applied in a rudimentary way as each corner edge had to move in two directions simultaneously.

When combined the actual incline ratio of each leg is height 2 to horizontal sqrt2.

If a building has corners which are inclined symmetrically to a central point of known height, the diagonal elevation can provide a slope ratio based on whole numbers. Had the designers been aware of the effectiveness of the Virtual Apex Method, they would have been able use this ratio to calculate systematically the exact outer corner position for each leg at certain heights using a reference scaffold and the centrepoint. A Virtual Apex height of 120m and a ratio of 4:3 would closely match that found. Reference heights could be every 4 metres and the first connection point at a height of 56 m would have outer corners 48m from the centrepoint. Had each leg's box section dimensions been based on a diagonal of 20m, their inner connection point would be 28m from the centrepoint and the horizontal cross-members connecting them 39.59m in length. (28Xsqrt2). If the calculation had been based on an Ancient Egyptian method using 99÷70 as a substitute for sqrt2 the dimension is 39.6.

(28÷4=7X9.9=39.6)

It is interesting to compare these simple methods with the profile dimensions to 4 decimal places (millimetres) resulting from Eiffel's plans and prepared in the days before electronic calculators.

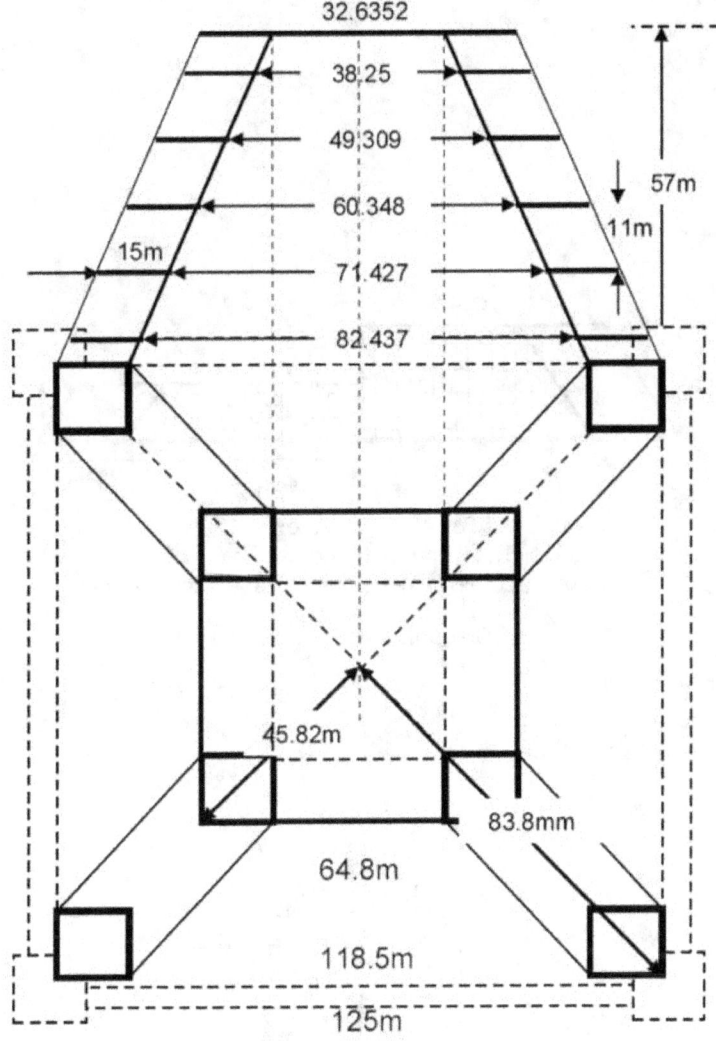

Eiffel Tower – Plan and Elevation
Dimensions at base and to first level at 57m
Taken from contemporary drawings

Legs slope to a 2:1 ratio in each direction

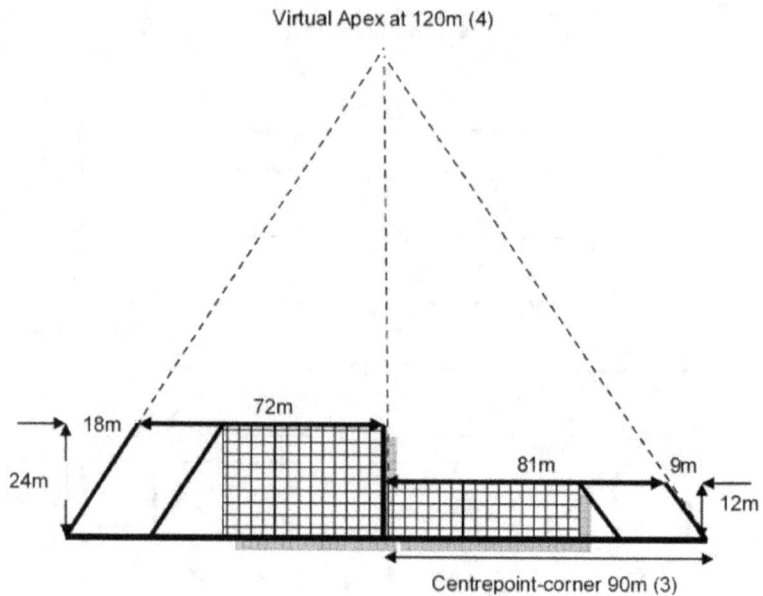

Eiffel Tower Virtual Apex at 120m
Height : Centrepoint-corner Ratio 4:3
Centrepoint-corner dimensions at Base, 12m and 24m

Diagonal Cross sections

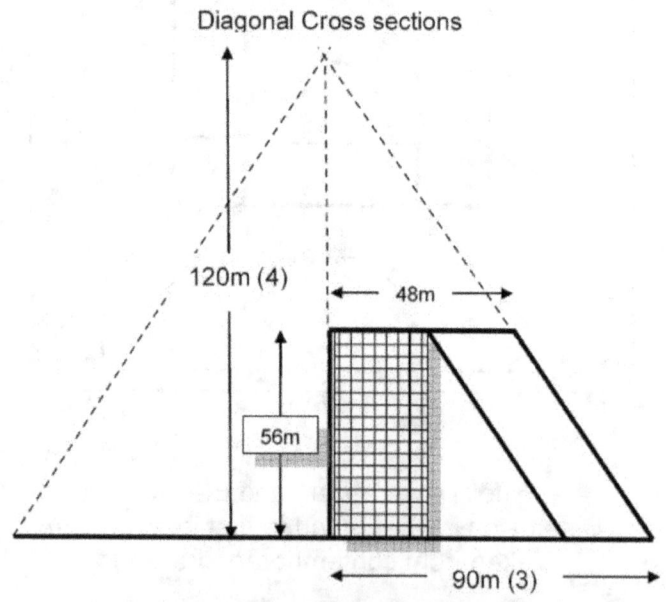

Eiffel Tower Virtual Apex at 120m
Height : Centrepoint-corner of Ratio 4:3
Centrepoint-corner dimension at a height of 56 metres
120 − 56 = 64/4 = 16X3 = 48

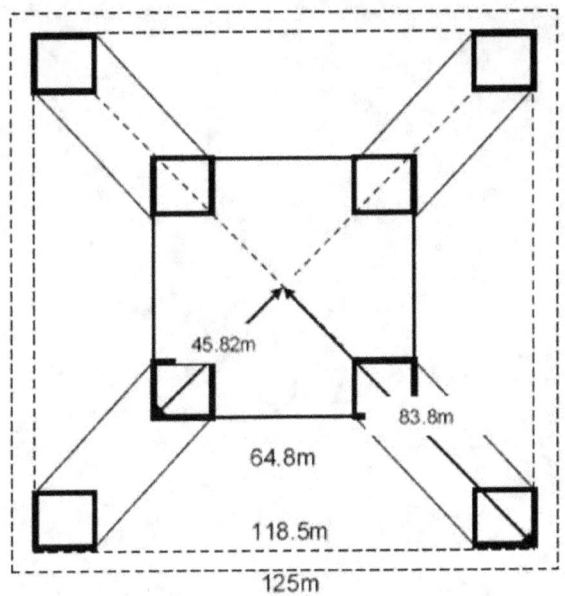

Eiffel Tower – actual dimensions at base and first level at 57m

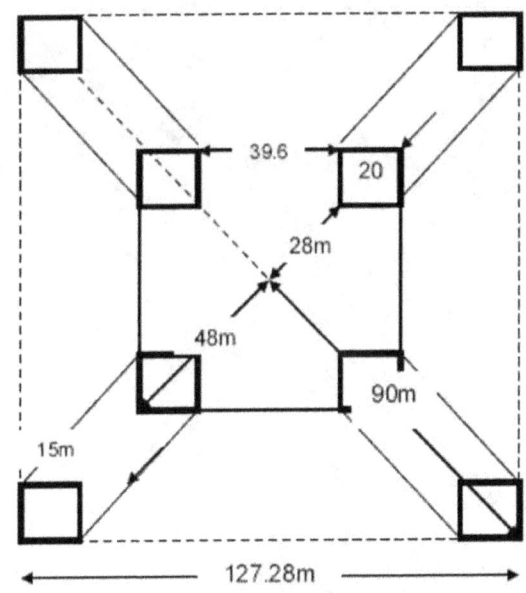

Dimensions resulting from applying a
Virtual Apex height:centrepoint-corner ratio of 4:3
Centrepoint-corner = 90m at the base and 48m at height 56m
A leg diameter of 20m gives whole number centrepoint-inner corner dimensions

References, Sources and Acknowledgements

Cyril Aldred: Egypt to the end of the Old Kingdom

Bancroft R M & Bancroft F J - Tall Chimney Construction - 1885 - Available Online

Somers Clarke and R. Engelbach: Ancient Egyptian Construction and Architecture

J.H.Cole: Egyptian Government Survey 1925

I.E.S. Edwards: The Pyramids of Egypt 1985

Michener James A. Caravans. 1963 ISBN 9780812969825

J.A.R. Legon: (www.legon.demon.co.uk)

Mark Lehner: The Complete Pyramids 1997

Petrie, W. M. Flinders. *The Pyramids and Temples of Gizeh*. 1st ed. London: Field and Tuer; New York: Scribner & Welford, 1883. Republished online at *The Pyramids and Temples of Gizeh Online*. Ed. Ronald Birdsall, 2003. Rev. August 27, 2014 <http://www.ronaldbirdsall.com/gizeh>

Hugh Thurston: Early Astronomy 1994 Revised April 1996 and www,dioi.org 2003 volume 13

Steven Haarck: Journal for the History of Astronomy 1984 volume 15

Star map from: www.fourmilab.ch/yoursky/

Wikipedia: Great Pyramid – Construction (Ramp Designs)

(www3.shropshire-cc.gov.uk/roots) - navvy workrates

Photographs courtesy of **Jon Bodsworth** (www.egyptarchive.co.uk)

Early photos and illustrations – **Charles Piazzi Smyth, John and Moreton Edgar, W. M. Flinders Petrie** and public domain

Eiffel Tower dimensions www.tour-eiffel.fr

Other illustrations and diagrams (except those otherwise credited) are by the Author

A time-lapse video of the construction process described here can be viewed on YouTube as: **'The Great Pyramid Construction Challenge'**

www.ingramcontent.com/pod-product-compliance
Lightning Source LLC
Chambersburg PA
CBHW071837200526
45169CB00020B/1646